DOWN
AND
OUT
IN NEW
ORLEANS

Studies in Transgression

Studies in Transgression
————————————————

Editor: David Brotherton

Founding Editor: Jock Young

The Studies in Transgression series will present a range of exciting new crime-related titles that offer an alternative to the mainstream, mostly positivistic approaches to social problems in the United States and beyond. The series will raise awareness of key crime-related issues and explore challenging research topics in an interdisciplinary way. Where possible, books in the series will allow the global voiceless to have their views heard, offering analyses of human subjects who have too often been marginalized and pathologized. Further, series authors will suggest ways to influence public policy. The editors welcome new as well as experienced authors who can write innovatively and accessibly. We anticipate that these books will appeal to those working within criminology, criminal justice, sociology, or related disciplines, as well as the educated public.

Terry Williams and Trevor B. Milton, *The Con Men: Hustling in New York City*, 2015

Christopher P. Dum, *Exiled in America: Life on the Margins in a Residential Motel*, 2016

PETER
MARINA

DOWN
AND
OUT
IN NEW
ORLEANS

TRANSGRESSIVE
LIVING IN THE
INFORMAL
ECONOMY

Columbia University Press
New York

Columbia University Press
Publishers Since 1893
New York   Chichester, West Sussex
cup.columbia.edu
Copyright © 2017 Columbia University Press
All rights reserved

Library of Congress Cataloging-in-Publication Data
{to come}

Columbia University Press books are printed on permanent
and durable acid-free paper.

Printed in the United States of America

Cover design: Lisa Hamm
Cover image: © Todd Norman

This book is dedicated to the city of New Orleans
and all those who call her home.

CONTENTS

FOREWORD

David Brotherton

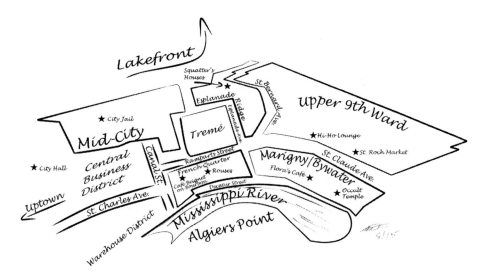

MORE THAN TEN YEARS AGO, a young man from the New School in New York City sat in my graduate class at the City University of New York and talked about the styles of youth in his hometown of New Orleans. He waxed lyrical about how these youths were not well understood and that the sociology of working-class young people usually focused on their misdeeds and the fears of society about teenage deviance and threat. What this young man was essentially describing was the process of moral panics, in which certain constituencies look for people to blame for society's shortcomings. The same young man then went on to argue that no one has really sufficiently studied the culture and vitality of his hometown and how this infamous setting needed to be explored by one of its own. Little did I know that some years later this young student would be the author of this book, which I very much welcome into Columbia University Press's Studies in Transgression series.

Peter Marina's innovative and refreshing sociological foray into the nether regions of New Orleans's social life, a little over a decade after the natural and human-made catastrophic events of Hurricane Katrina, is an ethnographic triumph by one of that city's native sons. Driven by his love of the city and its world-renowned culture, which celebrates both life and death with equal abandon, Marina embarks on a journey to understand whither his community in the face of some of the most brutal winds of change befalling any U.S. urban population in the current period.

This work establishes Marina as one of a new breed of "transgressive" sociologists, a social scientist who writes against the grain of staid and conformist orthodoxy. Dedicated to the fullest representation of our fellow human beings who exist "in the margins," he sees them not as cardboard-cutout Others, replete with "deviant" typological properties, but as dynamic and contradictory social agents who make the world, as Marx famously intoned, not in the conditions of their choosing "in self-selected circumstances." For Marina the subjects in his study are precisely the unsung folks who actually make the city what it is, who contribute to its survival and its ongoing vitality. However, Marina is not just another neutral observer casting a colonizing and patronizing gaze on the multitude but a reflexive participant in the creation of a world where the script is still being written. Along the way in his journey into New Orleans's "soul" he does his utmost to chart its progress— and he is always clear which side he is on.

As such he celebrates ingenuity over conformity, struggle over acceptance, resistance over adaptation, and creativity over the mundane. The result is a work of sociological discovery, an ethnographic tour de force, and a clarion call for the restless among us to follow our hearts and imaginations rather than our checkbooks and careerist pretensions as we go about making life. While there is pain and frustration in the subcultures he enters, there is also joy and willfulness and a healthy dose of dissent. New Orleans is threatened, yes, not simply by climate change and the waters of the great Mississippi and the Gulf but by the capitalist logic of accumulation that commodifies, reifies, and privatizes everything and everyone in its path. "Now is the time to save

New Orleans," says Marina. "It's time to save it from the new monsters of modernity that wreak havoc on our land and culture. Only the people of the city can prevent New Orleans from becoming down and out."

Inspired by the work of George Orwell in his best-selling chronicle of 1930s poverty, *Down and Out in Paris and London*, Marina uses his vast store of insider knowledge to excavate the city's "underbelly" and introduce the reader to a scintillating array of cultural happenings, characters, and traditions that read almost as the seductive peregrinations of a travel writer in full flow as much as the observations and reflections of a trained social scientist. But that is Marina's forte, just as it was Orwell's, the ability to use the pen to draw the reader into the fantastical workings of social life, which on the surface seem perhaps predictable and banal but underneath bubble with a certain wondrousness and unpredictability that, as he says, gives the city its "soul."

In his approach to research, Marina pays particular attention to Orwell's extraordinary eye for detail, ear for language, and sensitivity to the intricate dramas of the everyday. Similar to Orwell, who painstakingly portrayed the microworlds experienced firsthand during his seven months as a *plongeur* in Paris and as a hobo in and around London, Marina goes about the tasks of documenting and revealing life as it is lived in real time through hanging, befriending, and toiling with a cast of characters he simply encounters on the way. And like Orwell, he subsists only on the money he earns as a bartender, Airbnb go-between, and street performer, carefully noting how he learns and plies his trades, making it from one day to the next. In a generalized society still resolutely strapped to the cash nexus, where the spaces to be ourselves are increasingly restricted, policed, and surveilled, Marina points us to the possibilities of living in between the grids of work and play, in an environment that still has that capacity and commitment to welcome and absorb the seeker, the traveler, and the striver.

Marina gained this knowledge not simply from reading books but more importantly through a lifetime of experience that saw him attached at the hip to the city's fate. He was born and raised on Odin and Spain streets in Gentilly, on the edge of the French Quarter. The son of Cuban immigrants, with a father who for many years directed narcotics enforcement in the New Orleans Police Department, he went on to graduate from the city's school system, gained an undergraduate sociology degree from the University of

New Orleans, and later became a social studies teacher in one of the city's public high schools. It was only when Marina came to New York for graduate school that he left the city, and it has always been his dream to return to his roots and use his skills and training to shine a light on this piece of Americana that has so conditioned his life.

But it should not be forgotten that the book is set in a city that has endured enormous stresses and strains, economically, politically, and socially. While many point to New Orleans's steady recovery in the post-Katrina era, it has also suffered from disastrous cuts in public investment at the hands of an uncompromising governor whose aversion to taxes virtually bankrupted the state. In addition, the city had an egregiously corrupt political leadership, with its former mayor receiving a ten-year prison sentence for bribery and corruption—surprisingly, the first New Orleans mayor to face charges based on his time in office. Meanwhile violence, poverty, and inequality have continued apace. In 2015 the homicide rate was 47 per 100,000, second only to St. Louis for a city of its size. Roughly 28 percent of the general population is poor; fully 44 percent of the black population has been mired in poverty for the past two decades. Meanwhile, income distribution is so skewed that the top 5 percent make eighteen times more than the bottom 20 percent, the second-worst record of one hundred metropolitan areas in the nation. Of course, the problems of race and class were widely exposed and exacerbated by the flooding waters of the Gulf, when the city was evacuated of thousands of vulnerable black residents who were then not provided the wherewithal to return. The result was a drop in their proportion of the population from 67 to 59 percent; quite naturally, charges of racial cleansing continue to be leveled. As Marina notes, much has happened since the day the levee broke, and that story needs to be told.

Marina knows many of the city's various guises and has his antennae finely tuned to the whereabouts of the newest additions. On various occasions we are taken on a tour of its night life, to the jazz clubs and pop-up jams in and around the French Quarter, to the vast assortment of bars and their signature alcoholic offerings, to the restaurants and cafés boasting endless Creole concoctions, in which oysters, crawfish, crab, and shrimp always seem to feature. He also navigates with ease the patchwork quilt of highly complex residential neighborhoods, and he is well acquainted with their colorful local histories. And last but not least, he has that native's grasp and appreciation of the changing social geography of a place that still feels more like a slice of the Caribbean than an intrinsic part of the U.S. mainland.

It is through his carefully recorded and interpreted travails and urban adventures that the reader is able to feel and taste the layered culture of the city. Marina describes a series of "scenes," the settings in which occultists, Satanists, squatters, buskers, hustlers, and various types of street performer mount the stage and strut their stuff. To Marina, these constitute the social stuff that makes New Orleans unique, that original place of deviance where the normative rules of Southern propriety are broken as a matter of course and carnivalesque behavior is encouraged and socially reproduced. Only here can that wonderful hybrid subject, the Mardi Gras Indian, get to parade his or her identity in open defiance of racial markers and cultural conformity.

But real sociological questions constantly guide Marina the researcher. How does one negotiate life on a few dollars and cents eked from those formal and informal economic opportunities that are the lot for so many lower-class New Orleanians and spirited wanderers born into or avidly searching for other kinds of America? What social networks do we become part of in this community of seekers and misfits? How does one live outside the corporate imagination that dictates our daily routines in the rest of the United States? Who are these urban nomads making New Orleans their new sanctuary? How does the settled, indigenous population take to those who have the means to be mobile and the spirit to spread their wings? And will New Orleans go the same way as San Francisco, just another urban Disneyland film set where the well-heeled can practice their narcissistic moves and sate their voyeuristic appetites? Marina pursues all of these lines of inquiry and many more, and by the end of his personal and scholarly journey you have a feeling that this project has given him the taste to explore other subterranean milieux that late modern America is busily churning up.

In a United States where less than half of the population has full-time jobs (during a so-called recovery), producing a new social class, the precariat; where power and wealth are so heavily concentrated among the top 0.1 percent; and where affordable housing is a relic of the past, it is difficult to see how these bands of rootless ones do not become the norm rather than the exception. As Jock Young, a cofounder of this book series and a great sociological and criminological influence on the thinking of Marina, saw, in this era, capitalist cultural inclusion, especially around notions of the American

Dream, is often met by social exclusion, leaving many to go in search of their place in a world that is confusing, chaotic, and often hostile. Nonetheless, as Marina reminds us, never underestimate that spirit to survive and attain a semblance of the freedom supposedly promised us in the Constitution. And it is this spirit above all that is alive and well in New Orleans and that no rising waters can drown nor gentrification co-opt.

Thus, it is ten years since Peter made his promise to return.

ACKNOWLEDGMENTS

THIS BOOK was both a sociological endeavor and personal journey to discover social life as it exists in the cultural underbelly of the city. Although the main character is New Orleans and those that live within her, this book says something about the rapidly transforming modern city today and the lives of its inhabitants. As the late postmodern metropolis undergoes rapid social transformation, urbanites respond in creative ways, carrying out increasingly transgressive lives that, in turn, shape this great human experiment we call the city.

As for myself, I love the city, its endless contradictions, its spaces for self-realization and actualization, and its cracks and crevices that allow us to carve out both magical and mundane lives, whatever your fancy. The city entraps us as well as frees us, and freedom, whatever that means exactly, seems something worth fancying. The city allows urban dwellers the ability to push the boundaries of the absurd; it allows us both to conceal and reveal; it makes our lies true and our truths illusions. We find our soul in the city, for better or worse. The city exposes our vulnerabilities and allows us to revel in our mess while somehow reclaiming dignity to all that is our pure selves, however tainted we may be.

New Orleans is my lover, and my lover torments me so good. She is my home, though sometimes a home that seems far away. Her caress brings a celestial joy, but sometimes her touch also bites. The city destroys and saves us. Humankind lost so much in our creation of the modern metropolis, all the security, safety, belonging, attachment, and all the walls of protection from the lurking anomie that waits for us all. But it gave us the ability to pursue something more, the freedom to fail recklessly on our journey of self-discovery and quest to find out exactly what this thing called "human" means. Thanks to the city, you wretched beautiful beast. Thanks to New Orleans, you torturous lover.

The writing of this book took place in New Orleans and St. Paul, near both ends of the Mississippi River. Writing in coffee shops allows the sociologist to concentrate on writing prose and to observe social life. The trick is finding good coffee shops. Thanks to Fair Grinds Coffee in New Orleans and Nina's Coffee Café in St. Paul for the ability to write and observe over a cup of coffee.

The New Orleans journalist and movie critic for the *Times Picayune*, Mike Scott, used his excellent editorial skills and knowledge of New Orleans to offer suggestions, advice, and comments throughout every chapter. His skillful eye and insightful suggestions for this book helped shaped it into its final product. Thanks, brother, for everything. Who dat.

Thanks to Shannon Monaghan, Kesha Young, Shane Sayers, Josh Sharp, Anastasia, Eric Odditorium, Stumps the Clown, Tim the Gold Man, and so many others who offered their time, thoughts, and experiences throughout the research process. I would like to thank Shannon for keeping the legacy of traveling alive.

Special thanks to Noelani Musicaro, a woman of great kindness indeed. It's not often that research participants become genuine friends. Josh Sharp is my friend. Thanks to Pandora and the Mudlark Public Theatre.

May the brass bands, Mardi Gras Indians, and all the second lines live on forever.

Special thanks to my mentors and dear friends David Brotherton and Jock Young. The spirit of the late Jock Young forever lives on. Thanks to the critical criminologist Jayne Mooney. Thanks to my good friend David Gladstone, who continues to inspire me. Let's continue to see the world before it goes. Terry Williams continues to inspire my drive as an ethnographer of the city. And thanks to Bill and Barb Zollweg. Bill is one of the greatest teachers of sociology I have ever known.

Thanks to my beautiful Boston Jew rebel Barry Spunt and his New York City heroin research. Thanks to my wild-haired brother from another mother, Danny Kessler; few men in the world possess the wisdom and gentle sensitivity to the human condition as he. Thanks to my wonderfully mad friend Louis Kontos.

Bobby Kemp is my dear friend. We grew up in New Orleans together and continue to share many adventures in the city. Thank you, brother, for your friendship. Jimmy Lightfoot is a true friend who grew up with me in New Orleans. Jimmy is a mad one, and I love him for it.

Thanks Mom and Pop, respectively, Elena Perez Lopez Marina and Pedro Carlos Marina, for all your support and love. Thanks to Ben, Sam, and Madison Scott. Thanks to Paul, Elena, Roxanne, Lagniappe, Lorelei, Jason, Brandon, and Hailey Marina. May we keep the spirit of nuestras abuelas, Benita Marina Giniebra (a.k.a. Mima) and Silvia Lopez Ventura de Perez (a.k.a. Gagi), alive forever.

Thanks to the sociology department at the University of New Orleans for providing me an office to interview some of the research participants. Thanks to UW–La Crosse.

Many thanks to the outstanding staff of Columbia University Press and its wonderful editors, especially Jennifer Perillo, Stephen Wesley, and Robert Fellman, for bringing this book to press. It was a joy working with such a dedicated team of professionals producing top-notch scholarship.

Thanks to the spirit of George Orwell; your insights still guide us.

The last words of this book were written in the city of New Orleans. To all the people of New Orleans living in the city and throughout the world, stay strong—We Dat.

Peter Marina
New Orleans, 2016

DOWN
AND
OUT
IN NEW
ORLEANS

CHAPTER 1

NEW ORLEANS

Romancing the City of Sin and Resistance

The Rue du Coq'Or, Paris, seven in the morning. A succession of furious, choking yells from the street. Madame Monce, who kept a little hotel opposite mine, had come out to the pavement to address a lodger on the third floor. Her bare feet were stuck into sabots and her grey hair was streaming down.

—George Orwell, *Down and Out in Paris and London*

RUE TOULOUSE, NEW ORLEANS, five in the morning. Sounds of late-night debauchery permeate the early-morning slumber on piss-stained streets. Shane, who rents a run-down—and barely legal—shotgun hostel in the city's Seventh Ward,[1] emerges from the room with a cigarette in his mouth and a guitar strapped to his shoulders to talk with a short-term renter from Anywhere Else, USA. Shane's sweat sticks to his face while he peers at his tenant, still staggering from a bellyful of the local Abita beer.

"You play music?" Shane asks, already knowing the answer.[2] Almost everybody first moving to this mosquito-infested swamp city bojangles some type of tune. The tenant replies, "I get on," knowing damn well Shane might be a connection to a connection within and among the urban tribes of New Orleans. Urban subcultures of musicians, artists, anarchists, druggies, and intellectuals from a wide variety of creative endeavors conglomerate in the Crescent City, gravitating to one another in a shared pursuit of drugs, sex, music, art, and enlightenment. Shane says, "Well you know what to do, Monday night, back of Molly's, eight o'clock. A group of us do our thing." Just like that, we get another member of the underground urban scene.

Stories captured in this book—from that of "New Orleans Saved Me" Shane to those of a romantic Frenchmen Street poet, a cemetery-jumping occultist, a freak-show sword swallower, and dozens of others—capture the worlds of similarly willful fringe dwellers. These are souls who exist—with varying degrees of success—within the many subcultures of this new bohemia. New Orleans, and its central heart—the Vieux Carré, better known to many as the French Quarter—has been attracting bohemian types for over a century.[3] And while they come in all different shapes, sizes, and flavors, they share a unified desire: to carve out creative lives that transcend the ordinary. This book is an homage both to Orwell and to the resilient "accidental" city of New Orleans,[4] viewing the city from the prism of downtrodden 1920s Paris to showcase the fascinating and uneven journeys of late-modern New Orleans urban dwellers and tribes living on the periphery of city life. It reveals the complexities of social life within the cultural mecca that produced jazz and brass-band funk. What follows is a journey into the beating heart of an unapologetically contradictory city, one that can at once be stagnant and vibrant, exhibitionistic and mysterious, celebratory and treacherous, rich and poor—and eternally fascinating because of it all. The Crescent City we will visit down near the mouth of the Mississippi River isn't the postcard-ready New Orleans highlighted in the *Lonely Planets* and *Zagats* of the word. It is the Moulin Rouge by way of Baton Rouge.

More than eighty years have passed since George Orwell wrote *Down and Out in Paris and London*, which intimately described the experiences of poverty in the great modern metropolises of his time. The literary critic and philosopher Walter Benjamin may have captured 1920s Parisian social life as a *flâneur* strolling through the city as a connoisseur of the street,[5] but Orwell lived and worked in the urban underbelly beneath the iron-and-glass-covered arcades.[6] The interwar Paris of Benjamin's and Orwell's writing attracted a creative community of bohemian artists and intellectuals who frequented coffee shops and participated in lively nightlife. In bourgeois Paris, creative artists and intellectuals—the heart of Stein's "lost generation"—attempted to find meaning in a world that had proved worthy of distrust and cynicism. The Paris of the 1920s became the breeding ground of Hemingway's battered but not lost resilience and Fitzgerald's disillusionment with extravagance and the failed promises of Western society.[7] It attracted migrating expatriate urban dwellers searching for meaning in an irrevocably damaged world. And beneath the surface of Benjamin's bourgeois Paris, Orwell depicted life on the fringes of the city's Latin Quarter as he stood in breadlines, worked lowly restaurant jobs as a *plongeur*,[8] experienced near-destitution, and traversed the unseen and often hidden spaces of the city. During this same period, many artists and intellectuals were also arriving in New Orleans, making that city—the cradle of jazz—a Paris on the Mississippi.

While the uptown neighborhoods attracted the bourgeoisie, who lived along St. Charles Avenue, the grit of the Vieux Carré just downriver attracted, à la Montmartre, artists and intellectuals. "Dixie" Bohemia[9]—where the beer was cheap and rents were low, women entertained for agreeable prices, and tolerant attitudes for the madness of creativity prevailed—became fertile ground for alternative and resistant lifestyles.[10] New Orleans in the 1920s, the Vieux Carré in particular, became a literary hotspot, beginning with the journal *The Double Dealer*, which published the works of, among others, Sherwood Anderson, William Faulkner, and Ernest Hemingway—all literary giants who either lived in New Orleans or who had spent considerable time there during this period.[11] It was where William Faulkner and William Spratling wrote the pamphlet "Sherwood Anderson and Other Famous Creoles," and it was a place composed mainly of migrant bohemians, writers, preservationists, intellectuals, and artists, all participating in the social and cultural life of what was then "surely the most civilized spot in America."[12] It's the place where Tennessee Williams has Stanley shout "Stella" into the humid, uneasy air,[13] where Louis Armstrong Muskrat Rambles the Basin Street Blues of Storyville, where F. Scott Fitzgerald drinks sazeracs at the Red Room in the Roosevelt Hotel,[14] and where Sherwood Anderson writes about sexual freedom in *Dark Laughter*, based on his experiences in New Orleans while living in the historic Pontalba Apartments of Jackson Square.[15]

Fast-forward a century, to post-Katrina New Orleans. It still attracts a flow of modern bohemian transplants, all migrating to the city to become part of its distinct culture and in search of cultural authenticity, artistic expression, self-actualization, identity transformation, and alternative lifestyles. Just as in

1920s Paris, beneath this bohemian surface lives a huge urban and mostly unacknowledged class of willful outsiders, people who exist on the edges of the postindustrial tourist economy. In this book, borrowing a page from Orwell, I pull back the veil to reveal the less-seen spaces of New Orleans, intimately depicting the social life of the new creative urban dweller living in or near poverty. As I describe in chapter 3, I lived on limited funds, like Orwell did when he first became down and out in Paris; I worked odd jobs in the city for ten weeks in New Orleans, as Orwell did in Paris, to try to re-create Orwell's work as closely as possible in the modern day.

I do not merely *study* the city's fringes; I live on them. I don't merely interview the city's new bohemians; I live among them, live as one of them. I don't settle for examining the city's modern underbelly. I creep and crawl through it, working menial jobs, scrounging for enough to eat, living among the urban tribes—becoming part of them—and hoping to survive.

I walk on glass as "Cuban Pete the Clown," participating in sideshow freak shows. I pantomime on the streets outside the Superdome before Saints games, busk poetry on Frenchmen Street, bartend on Bourbon Street, clean Airbnb apartments in Faubourg Marigny, trip with occultists in Barataria swamps, break into cemeteries for underground Satanic rituals, attend informal burlesque shows, sleep in homeless shelters, and squat in abandoned buildings.

Using Orwell's *Down and Out in Paris and London* as a rough blueprint, this book reproduces, as closely as possible, the conditions Orwell faced while down—but not completely out—in late 1920s Paris. The goal: to tell the story of postmillennium New Orleans and its culture of creative degenerates, vagabonds, artists, hustlers, lowlifes, transients, grifters, intellectuals, musicians, informal educators, druggies, skells, writers, gutter punks, goths, nihilists, and existentialists who exist beneath the radar. This book is a transgressive sociology about the world of Lucky Dog vendors, seekers of the occult, brass-band musicians, clowns with facial tattoos, burlesque dancers, buskers, hustlers, freak-show performers, beggars, hipsters, and white middle-class bohemians rejecting humdrum conventional life. We enter into the sights, sounds, smells, tastes, textures, and moods of real events among romantic nihilists living on the edge in a city that—because of its geography and vanishing marshlands—is also living a precarious existence.

Welcome to New Orleans. Welcome to the city of unapologetic sin, orgasmic "rattle-your-bones" romance, and stubborn resistance. Take a peek into the heart of New Orleans's nouveau bohemia. This is Frenchmen Street.

THE BLUE NILE ON FRENCHMEN STREET:
STOOGES BRASS BAND

"Marigny strolling" in the deep, dense, and heavily moonlit night along Frenchmen Street, just behind the Vieux Carré. Stumbling and bumbling and rumbling, it's New Orleans city-bouncing from d.b.a. to the Blue Nile music venues. It's all smoke and beer and whiskey and sex and sax and funk and trombone. The band starts its third piece of the first set. It's Saturday night; everything clicking just right. Big-man sousaphonist blows left and right, up and down, pumping knees marching style while blowing the hell out the brass. The crowd, hot and high and frenzied, swings forward, backward, left, right, it does not matter. Trombonist with cheeks puffing out like a blowfish about to explode. Drummer desperately tries to maintain pace, wildly beating on Kerouac's "rolling crash of butt-scarred drums."[16] Trombone man puts down his brass to shout, "You got to wind it up" and something about Michael Buck; the crowd cares less about the actual words because it's only the beat that means a damn thing.

The drummer plays a series of hits, a steady quarter-note conduit to a living breathing culture, like the Ogou giving passage to the spirit world of gods and their lambs. Mobs of revelers balancing beers and cocktails spilling all over the place; a smoke hangs from the right side of my lips, while a pretty one twirls like a well-balanced top from my left hand.

And it's not about being cool or dancing or the repetitive lyrics or the presentation of self or feeling as one collective connected to that proverbial something larger. There's also a sense of the sadness of the culture, but you have to work hard to become aware of it. You battle with the culture while it rejects you and your arrogant sense of self-importance. Arrogance is lost in the screaming trombones and beating drums; so is all sense of some essence of self or claim to authenticity or thoughts that you matter or illusions that you are part of something more. No one here tries to rescue individual subjectivity from cannibalizing objective culture. It's about realizing the simultaneous destruction of the self to save the self, the creation of culture through resisting it, saving the world by realizing it needs to be obliterated. It's about endless destruction to satisfy the human need for self-realization. The drummer beats and the trombonist trombones to funk and splash and shrieks, and no one knows yet that there is something here to get. It's as if with the music makers teasing out that moment of it, the first wave will know, and the knowing will know they know, and they will move in a knowing way until it spreads and begins to respond. A cultural dance to commence between the makers and receivers, this dialectical creation of music is a passageway to all that is frightening and history and myth and reality. Here in down-and-dirty NOLA we reclaim reified culture and give it our own human identity.

The mood suddenly changes when a sweat-drenched man with a long white T-shirt and baggy, saggy jeans demands that the frenzy stop, if only for a moment. He's got a story to tell. It's about his friend, a guy named Shotgun Joe Williams.[17]

The police killed Shotgun Joe Williams. They shot him dead.

The people don't know what to do but listen and try to feel what the man at the microphone says as he evokes his sadness and love for his friend now dead. And no one really knows except knowing that this has something to do with the culture and the music that is the voice of that culture. It gives new meaning to culture not as some unchanging noun handing off knowledge from one generation to another but as a verb that won't accept dying quietly in the unfriendly night. Culture here in New Orleans slaps you right in the face, like the hot sauce that squirts in your eye from the sausage in your jambalaya. But if you look for it, even in the music, if you purposely look for it, it's as gone as the lover when the front door opens and the back door slams. It's a subtle reckoning, sometimes a feeble whimper, other times a collective sigh, still other times the collective roar of humans who say, "Yeah, you right." It is collective in its pure drive of individual human will. They ain't willing together, mind you. They willing independently. But they willing all the same. And that is the culture. You get into the culture once you finally reject the city, and only then will the culture truly allow you into its labyrinth. The roaring and dancing and kissing and fighting, it's all isolated individuals collectively responding in creative ways to their shared structural circumstances. Shotgun said—"yeah, you right"—he was part of New Orleans, and the police killed him dead. But you can't kill culture and you can't kill Shotgun, and the band man gonna make damn sure of that. He says:

> He got killed by the New Orleans Police Department one day. The police stopped him as he was getting out of his car. The police shot him seventeen times in full view of residents in the city's Tremé neighborhood in New Orleans. And when we got to the scene of the crime and we asked the police officers what happened, like what he did, what was wrong, they told us they owed us no explanation. They tell us that today. This case was still not solved, so this song, "Why Did They Kill Him," and we do it every time we do a show y'all, so, rest in peace.

He don't wait for reaction, just goes groove but with matchstick fire energy as vocals bellow out like Paul Revere on his proverbial horse, "Oh Whyyyyyyyyyyyyyyyyyyyyy?"

And the people are rocking. They don't know Shotgun, but at this very moment they all wanna shoot the bastard cop who shot 'im.

And that last line, nobody cares cause everybody knows and the movement of the people and the movement of the band synchronize not because the band says so or because people try but only because they getting the movements of the culture that lies dormant in the deepest of souls.

And the crowd, lily pasty white to charcoal black and every shade in between, stick up their middle fingers and collectively scream with fingers in the air: "We say fuck the po-po."

NEW ORLEANS CULTURE AS SIN, ROMANCE, AND RESISTANCE

The city is a site of cultural creativity. It is where urban dwellers carve their subjective imprint on the hard spaces of urban design and architecture. New Orleans is a place where a Rabanesque image of the soft city—the human, flexible, malleable, and fluid city—springs to life.[18] New Orleans culture is as thick as the humidity; it weighs its people down with the intense burdens of the city's dark past while simultaneously lifting them to a transcendental, intoxicating awareness of its colorful and resilient history.

New Orleans is a unique blend of contradictions. It's simultaneously a shit-hole and a place of mysterious charm, a great display of extravagant wealth and dirt poverty, dog-tired and magically alive, empty despair and bursting possibility, static and kinetic, notoriously corrupt and courageously honest, otherworldly yet in this world, cutthroat violent and selflessly friendly, hostile and peaceful, ordinary and magical, naïvely innocent and wickedly tainted, subversive and submissive, highly resilient and passively succumbing.

The culture invents words and sounds that require new vocabularies of style. Its people "make groceries," walk on "banquettes," cross "neutral grounds," expect "lagniappe,"[19] say "who dat?" eat po-boys, ask "how ya' momma and dem?," inquire "where y'at?," and exclaim "laissez les bons temps rouler" in the face of an uncertain future and a precarious existence. "Mak-ing groceries" instead of merely "buying" them linguistically reflects active participation and involvement in the most mundane activities. The people sing when they talk. The culture is as diverse as its gumbo—creative, eccentric personalities living together while maintaining their unique individuality—and its people retain their subjective essence, refusing to become a homoge-neous mass.

New Orleans is a highly reflexive culture, and it is constantly aware of itself. New Orleanians drink to celebrate their sense of being connected to something much larger than themselves; simultaneously, they sense their iso-lation from the outside world. They dance to brass and jazz bands to embrace life, knowing it can all fall apart with little warning. They embrace "the moment," knowing that one day it will all be gone. They indulge in cuisine as unique as its city—seafood gumbo, crawfish étouffée, shrimp creole, rabbit jambalaya, red beans and rice, raw oysters, fried catfish, turtle soup, muffu-lettas, spicy boiled crawfish, barbecue shrimp—to dull the sadness of the constant threat of extinction from both manmade and natural disasters. External happiness masks internal suffering. To be from New Orleans is to long for it even while in it. New Orleanians in exile wax poetic for their city.[20] They drink to forget, which makes them remember better. They revel in the precarious uncertainty of their lives, a future destined to falter, to feel life—to feel how deep the word "alive" actually means. Their mystery is as real, perhaps more real, than the conventional outside world. It's a place just as unique in death as it is in life: Mourners second line from the cemetery, refusing to accept death, instead celebrating life. Funeral homes in New

Orleans's black communities sell package deals that include memorial T-shirts that celebrate the dead along with a church service and second-line parade.[21]

In the City That Care Forgot, the people dance to revel in the human condition.[22] The people work jobs to make time for real life, best expressed through the fests—Jazz Fest, French Quarter Fest, Voodoo Fest,[23] Satchmo Fest, Zydeco Fest, Essence Fest, Decadence Fest, Mardi Gras Fest—where they release the collective sigh of their oppression-resistant creature. There even exist festivals to celebrate the city's unique cuisine, including the Mirliton Fest, Oak Street Po-boy Fest, Creole Tomato Fest, Seafood Fest, and Tremé Creole Gumbo Fest. And, of course, there's the Tales of the Cocktail Fest to celebrate the city's fine drinking traditions. New Orleanians go to coffee shops—Sacred Grinds, P.J.'s, Morning Call—to discuss local music, politics, existentialism, and theology. But mostly they talk about their city. Stepping outside from a local café to take a break from writing, I overheard two transplanted locals waxing poetic about the city. "Everywhere is fake kindness," one says. "New Orleans is a forgiving kindness."

In New Orleans the locals complain about the endless corruption, the proposals to tear down their housing projects, the devastation of gentrification, the rising housing and rental prices, the crooked politicians, the old money, Hurricane Katrina, the new transplants affecting the culture, and the shocking murder rate. (New Orleans is often ranked as the deadliest city in the United States.) But if an outsider dares to point out these same gaping flaws, that Southern charm fast dissipates. Only insiders can rag the city. Outsiders ain't earned dat right. Dey don't know where dey at.

New Orleans residents constantly repeat the phrase "only in New Orleans," reveling in the many eccentric and peculiar characters that produce the fascinating visual scenes of the public sphere. New Orleanians deeply feel their sense of being. They are in the here-and-now but transcend their immediate realities to connect deeply and intimately to the larger culture of the city.

New Orleans culture is like an onion: The more you peel, the more layers get uncovered, the more complex it gets. The people see the taste of the oysters, smell the sounds of brass bands, feel the sights of their second lines, and hear the visual movements of the dancing revelers. The music scene captures the collective nomic isolation of the people, who celebrate their cultural products through their actions.[24] They celebrate as a collective and, in doing so, produce the cultures of which they are a part, yet they feel the individual isolation that exists within that same culture. They find belonging in their collective isolation. To be from New Orleans is to be part of something larger, connected to something unique, while finding comfort in loneliness.

New Orleans culture is a verb; its people understand and see their part in shaping it with their own subjective unique human imprints. Existing in this culture, breathing in its cultural aroma, affects its denizens just as it does the gumbo and French bread they eat. And, in turn, they simultaneously affect the culture.

The Tulane geographer Richard Campanella admits his need to use other languages when explaining the local sense of what it means to be a New Orleanian: "It's the *tout ensemble* of *laissez faire* plus *carpe diem* with a little *joie de vivre* thrown in as *lagniappe*." He stresses the importance of geography in explaining how "the live oaks, the heat, the humidity, the sinking soils, the termite-infested shotgun houses, the tragedies, the triumphs, the uncertain future" serve as essential ingredients in intimately connecting New Orleans culture and its people to the physical and sociohistorical environment. Campanella declares: "You've got to live here—you've got to *cast your lot here*—to lay unconditional, nonfigurative claim to the demonym 'New Orleanian.' It's not just a state of mind. It's a state of place."[25]

Yet the city—and its mental life as a state of place—has changed in crucial ways in the years following 2005's Hurricane Katrina: an accelerated gentrification process (especially in the Bywater, Tremé, and the Lower Ninth Ward), a sharp increase in poverty, growing inequality of epic proportions, an epidemic of shocking violent crime, increased racial hypersegregation, influxes of Hispanic immigrants, and a sharp increase in un- and underemployment. In many pockets of the city, from the Seventh Ward to Hollygrove, the rates of poverty, illiteracy, and infant mortality rival some of the poorest countries on Earth, seriously challenging the myth of first- and third-world distinctions. These changes have been accompanied by an increased distrust of police and public officials, who have often done more harm than good. Some of these striking changes involve the regulation of the city's very culture, including the performance of music in the public space, under the auspices of Mayor Ray Nagin, the city's chief executive during Katrina and in the tumultuous years that immediately followed, and his successor, Mitch Landrieu.

The United States has pre- and post-9/11 worlds. New Orleans has pre- and post-Katrina worlds. The post-Katrina world is one of contradictions that involve, among other things, increasing gentrification and "yuppification" of many of the city's historic neighborhoods, shadowed by increased poverty and misery in some of its other neighborhoods. Even ten years after Hurricane Katrina, New Orleans continues to experience an increasing rate of migrant transplants, greatly changing the city's culture, visible in its oldest neighborhoods—from the Vieux Carré to the Faubourgs, including the Tremé, Marigny, and Bywater neighborhoods. Already a mecca for bohemian artists, writers, musicians, and other creative types before the hurricane, the post-Katrina transplants have revitalized an urban creative scene reminiscent of 1920s Paris and Jazz Age New Orleans—where jazz was born in Congo Square before spreading throughout the world. This book focuses specifically on the locals who were part of this creative scene before the storm and on the new transplants from all over the United States who settled in the city afterward.

DO YOU KNOW WHAT IT MEANS
TO MISS NEW ORLEANS?

Perhaps I should substitute "New Orleanians" and "they" with the collective pronoun "we." I know what it's like to miss New Orleans, to long for my city, its people, and its culture. I'm a native son of New Orleans, born on Canal Street near the Vieux Carré, schooled at De La Salle High School in the Garden District, college educated at the University of New Orleans on the Lakefront, and a resident of the working-class Gentilly and funky Mid-City neighborhoods. Over the years I have worked in about a dozen restaurants, from Dino's Pizza and Vazquez in Gentilly to Café Roma and Italian Pie in Mid-City, and I've also worked various other jobs in the tourist industry, from the Aquarium of the Americas at the foot of Canal to Daiquiris Shops on Decatur. I worked as a bartender at the Spider Web in Gentilly and at Parley's Tavern in Lakeview and as a bouncer at the ecstasy-riddled Club 735 near the corner of St. Ann and Bourbon. I was a teacher at my alma mater, De La Salle, as well as at John F. Kennedy, Warren Easton, and Abrahamson high schools, and I served as a visiting summer professor at the University of New Orleans, teaching criminology. While living in Brooklyn and earning my doctorate at the New School for Social Research in Manhattan, I longed for my New Orleans, soothing my expatriate soul with old Satchmo tunes and by streaming the only-in–New Orleans radio station WWOZ.

My formative years involved traveling to work with my father for police details at the McDonald's on Canal Street, where we would discuss sociological ideas while keeping the place safe for customers. Other times, we would talk while sitting on police cars during Carnival and watching the police ceremonially sweep Bourbon Street clear at the stroke of midnight Mardi Gras night. I picked magnolia flowers for my mother on the neutral ground of Elysian Fields—and I may have cried once when the Atlanta Falcons beat my beloved Saints. Memories of nighttime television include watching "horror host" Morgus the Magnificent and Frankie and Johnny's "Special Man" furniture commercials. My favorite childhood activities involved playing neutral ground football on St. Roch Avenue, baseball games at St. James Park near Milne's old "bad boys" home,[26] and pickup basketball at the old St. Raphael outdoor court near Elysian Fields. My family made groceries at such revered New Orleans institutions as K&B, Dorignac's, and Schwegmann's; devoured chocolate-covered "turtles" at McKenzies Pastry Shoppes; gobbled huge dressed fried shrimp and oyster po-boys ("so poor you can't afford the 'o' and the 'r'") at the Parkway Bakery;[27] and partook in the delights of Angelo Brocato's Ice Cream and Buddy D's "If ya ain't in Bucktown, you ain't at Deanies" seafood restaurant. Memories of New Orleans include soulful and reflective drives along Wisner Boulevard late at night, with a fog coming off the bayou. It's not an odd sight in this city to see a New Orleans native standing on the bank of a bayou feeding the nutria.

I've brawled in New Orleans bars, drunk away many nights within its most raucous scenes, danced seemingly endless nights in unrelenting hedonism, and had my heart broken when the Saints delivered yet another loss. I've marched with second lines and strutted with New Orleans's black Mardi Gras Indians, wandered the many aboveground cemeteries, and drank hundreds of cups of coffee at Fair Grinds, where I wrote my master's thesis. I broke hearts in New Orleans and had my heart broken; I've sung "Jock-A-Mo Fee Na Nay" at New Orleans life's joys and sufferings. I reminisced with old football coaches and UNO professors in the Mid-City watering holes Mick's

Tavern and Pal's Lounge; sipped sazeracs at the Roosevelt Room, where my grandfather worked during his first days after migrating from Havana; and once thought that all Americans ate red beans and rice on Mondays and enjoyed sucking crawfish heads.

I've had friends who were killed on the streets, lost students to murder charges, and seen countless lives destroyed in a city whose officials seem to care more for tourist dollars than for many of its longtime community members. I've experienced hurricanes, swum through the water pouring into the city, cleared it from my house, and rescued my sister trapped in the floods. I still make pilgrimages to the site of my childhood home on Odin and Spain streets, where Hurricane Katrina left nothing but a slab of concrete. Now even that is no longer there.

I danced at the goth and punk Whirling Dervish on Decatur Street, reveled at the Hi-Ho Lounge and Mimi's on the Marigny's second floor, got lit at d.b.a. and the Blue Nile, and tore down street signs in the Marigny during Mardi Gras. Straight up, I'm a New Orleanian who was once—and, deep down, still is and always will be—part of this fascinating, underground world of debauchery and sin, and I loved it all.

Just as for other New Orleanians, my ethnic identity stems from my belonging to the city. The city is more than just another place of birth. It is my mistress, my intimate lover.

I believe that my connection to New Orleans and identity with its culture complements my training and extensive research as a sociologist. I have the scientific tools to conduct objective research on city life on the urban fringes, and I can also offer unique insight into the passions, both of exulting joy and maddening anger, that New Orleanians have for their city.

"Who dat?" you ask? I'm dat. I am New Orleans. And she is me.

DO YOU KNOW WHAT IT MEANS TO *BE* NEW ORLEANS?

One cannot assimilate into the libidinous labyrinth of New Orleans culture through sheer will. First, one must let go of everything. All New Orleanians say, at least a few times in their life, "fuck New Orleans," which is why we also love it. Becoming New Orleanian is a classic contradiction; one must refuse

assimilation into the culture to become part of it. The collective consciousness of the city forms through radical individualism: Be what you are and not
like anything else. Intolerance to masked tolerance is what makes its people
so tolerant; one is expected to be different—so long as it is genuine. Becoming New Orleanian means becoming what you are, refusing to become like
others, being disquieted by sameness, and maintaining individuality.

Being a New Orleanian is more like an ethnicity, a status beyond race,
class, gender, and sexual identity—though these factors remain salient. It is a
way of being, a style of walk and talk, a specialized geocultural argot, a way
of listening beyond hearing, a performance of the body, a barrage of bodily
expressions and vocal sounds, a host of movements and utterances, an
identity with cultural signifiers that remind the city's inhabitants of its ethnic
pride. The fleur-de-lis, the symbol of New Orleans ethnic unity, covers the
city's public and private spaces, from the city streets to antiques in living
rooms to the tattooed skin over a resident's throbbing arteries. It is a city of
living symbols, and they intimately interact with the city's inhabitants. The
people recognize themselves in these symbols, and the symbols confirm the
people's sense of belonging to the city of longing.

The fleur-de-lis took on additional resonances after Hurricane Katrina.
Abandoned immediately after the levees were breached, the city learned a
lasting lesson about the illegitimacy of power and the failures of the nation-
state. The hurricane exposed the hypocrisy of the entire system, exposed the
cracks of hegemonic power, never again to be hidden away.

Today, the fleur-de-lis represents a collective unity of radical difference, a
sense of place and being, along with a belonging, perhaps better said, a right-
ful possession of the city's magic and a distinction from the rest of the country. Many of the spray-painted FEMA glyphs that once represented death and
abandonment on the facades of searched homes remain on some buildings, as proud indications of the inhabitants' resilience, strength, survival, and
rebirth. Though the hurricane caused suffering and loss, it had unanticipated
consequences that revealed individual strength, determination, resiliency,
and the power of human agency. Most of all, it awakened a people. To put it
in Marx's terms, New Orleanians are not only a class in itself but also a class
for itself.[28] The people continue to build the city, despite the efforts of the
elites. Dealing with government failures at every level, the people of the city

rebirthed New Orleans, and this remains a work in progress. As the people build, they create and transform both their identities and the culture, along with the New Orleans symbols that give material expression to that culture.

But New Orleanians are jealous lovers. The more they shape the city and its rebirth, the more they lay claim to the culture and the magic it possesses. Inauthentic outsiders seeking profit and gain—modern-day carpetbaggers—have no right to the culture, no matter what the city's policy and plans are. Those who seek entry to the culture find access through a desire for something more, for something meaningful, for something genuine.

ETHNOGRAPHIC ANARCHY ON THE URBAN FRINGES

This is a transgressive sociology that builds real and genuine relationships with people on the urban social fringes. To them, I'm not a researcher—although I never hid from them my reason for being among them. To them I am just another soul doing another creative thing.

The methodological approach taken throughout this research combines George Orwell's strategy of living and working down and out with C. Wright Mills's insights about avoiding rigid procedures and unimaginative rubrics. The goal is liberation from conventional procedures, the crutch of coding, and stifling orthodoxy, as Mills first advised and Jock Young later urged.[29] If this book captures the spirit of Orwell, Jock Young inspires the soul of it. Jock "yeah you right" Young.

My method of research requires establishing genuine relationships with people and becoming involved with their new cultures. It requires walking the streets in sync with the city beats, laughing and crying with others while sharing in the human condition. Ethnographic anarchy involves exposing vulnerabilities and taking risks, calling people out on their bullshit and admitting your own. Like David Brotherton with his banished Dominican deportees,[30] this involves not only seeing but feeling—deep-down-in-your-bones feeling—the world of others and taking into yourself their fears, joys, insecurities, heart, sufferings. If you do not cry, shout, get angry, feel moved, become emotional, and experience a wide variety of anxieties, fears, insecurities, and joys, you ain't doing it right.

WHAT HAPPENS NEXT?

Chapter 2 juxtaposes the hard and soft cities of New Orleans, where inhabitants carve unique biographies and cultures within the Spanish architecture and French-named streets. Describing the hard city and urban grids of neighborhoods provides the geographic context for the soft city that lies beneath the concrete jungle. Chapter 3 describes down-and-out urban living in New Orleans and compares it with Orwell's *Down and Out in Paris and London*. Chapter 4 delves into the vibrant and endlessly entertaining streets of the Vieux Carré, where buskers perform—and where a sociologist goes pantomiming and walks on glass in freak shows. Chapter 5 takes a closer look at the informal nocturnal economy of Frenchmen Street in Faubourg Marigny. Chapter 6 takes a journey into the world of urban camping, where young homeless travelers hunt for abandoned houses in the still storm-wrecked neighborhoods of the city. Chapter 7 ventures into the underground of witches, occultists, and Satanists of the Vieux Carré. Chapter 8 discusses the nuances of gentrification, showing how the old and wise city of New Orleans has fallen under the threat of large-scale structural changes. Chapter 9 describes the problems of gentrification and the unanticipated consequences of relative deprivation when hipsters attempt to create wonderlands in historic neighborhoods, potentially with deadly consequences. Chapter 10 discusses the brass bands and second lines of New Orleans and the world of those who produce the fascinating culture of the city, one that struggles to survive in the face of massive urban structural transformations. The conclusion depicts the city of New Orleans as it pushes toward an uncertain future and argues that resistance and struggle, subversion and transgression, ingenuity and creativity, and the ongoing fight against the powerful will perhaps save the city and our urbanized world.

CHAPTER 2

———

THE HARD AND SOFT CITY

A Portrait of New Orleans Neighborhoods
and Their Characters

The lodgers were a floating population, largely foreigners, who used to turn up without luggage, stay a week and then disappear again. They were of every trade—cobblers, bricklayers, stonemasons, navvies, students, prostitutes, rag-pickers. Some of them were fantastically poor. . . . There were eccentric characters in the hotel. The Paris slums are a gathering-place for eccentric people—people who have fallen into solitary, half-mad grooves of life and given up trying to be normal or decent. Poverty frees them from ordinary standards of behaviour, just as money frees people from work. Some of the lodgers in our hotel lived lives that were curious beyond words.

—George Orwell, *Down and Out in Paris and London*

THE TRANSPLANTS of New Orleans are a nomadic population, mostly American, who arrive with only a few dollars in their pockets and little more than the clothes on their backs. Some stay only days; others become absorbed into the city's underbelly. Many transplants float from place to place, usually with only a small bag of essential items, sometimes a guitar, some worn-out clothes, and a pack of smokes. They leave hometowns across the country—from Massachusetts and New York to California and Texas—to travel to the alluring Crescent City. Sometimes they stay for a while. Other times they stick around only for a few weeks before disappearing again. Some find friendships with fellow travelers, strangers that intersect with their path, divergent lives crossing in the long and turbulent journey of nomadic life. The travelers often work odd jobs as carpenter's assistants, waiters, and baristas, often after dropping out of college. Many come from families of wealth and privilege, and some are poor, sometimes incredibly poor, and all are seeking "it," that desire to rise above the ordinary, to find the magical end of the road among the "fellaheen" of the world.[1]

Shane, one of my contacts, was a drifter who realized early in life that he could not be part of the pack, the stupefied masses, the mainstreamers, the ordinary, the normals.[2] Perhaps there is nothing worse for the idealist than the idea of succumbing to the mundane. "I felt apart, like I was trying to agree with a game that I wasn't really wholeheartedly a part of," he says. "I didn't really want to fully participate in this game, but I was just doing it because [it was] the American dream. It's like, being popular, having friends, getting good grades, playing sports, going on to college, then, graduating, getting married, having children and then settling down, more or less." After wandering for years, he found his magic land. "New Orleans," he says, "saved my life."

The New Orleans faubourgs are a habitat for transgressive, eccentric, and often fascinating urban dwellers—people who have deliberately chosen to live as outsiders regardless of the social and personal cost. Some lead solitary lives. Others form tribes on the city's fringes and in the cultural nooks and crannies of the metropolis. They make their living in the informal economy and no longer abide by the mainstream rules of conformity. They might lead what some would describe as self-destructive lives, but they also find creative solutions to what they perceive as the problems of our late-modern era.

The people of New Orleans leave their marks on the malleable soft city, scratching their identities and places on its stubborn, resistant, but still pliable culture.[3] The down-and-out urban dwellers on the fringes of the city carve out their biographies in an attempt to rise above the ordinary in new and creative ways. Just about everybody wants to be somebody. Some find themselves, because of situations beyond the control of the private orbit of their lives, buried within hierarchical structures of power and ideology but reject their low positions within them. Others deliberately exclude themselves from institutional domination—from our stifling rubric-driven educational systems, from our illusions of democracy, from our oppressive economic

institutions, from the requirements to submit to the wills of our social superiors. While many normals live in New Orleans, there seems to be, though it
is certainly difficult to measure in any sure way, an extraordinary number of
strange and peculiar characters bursting with delightfully insane charm and
character.[4]

Here the question arises of how to distinguish oneself in a sea of such
creative characters. This is the stuff of Simmel's "strange eccentricities, metropolitan extravagances of self-distanciation, or caprise, of fastidiousness . . .
being a form of 'being different'—of making oneself noticeable," which
attempts to resurrect "the atrophy of individual culture" over and above "the
hypertrophy of objective culture."[5] One of the great struggles of New Orleans
residents is the task of developing a unique subjective self, one separate and
distinct from the mass of peculiar characters that externalize their subjectivities into the all-consuming, cannibalizing New Orleans objective culture.
That is no minor undertaking. It is, in essence, the attempt to save oneself
from the tragedy of modern culture—the annihilation of the subjective individual into the gumbo obscurity of New Orleans.

The people and the numerous subcultures and tribes of the city also
attempt to recover individual *geist* from an empty and watered-down com-
modity-driven culture that dominates the tourism industry. These willful
down-and-out urban dwellers have given up—or lost—the safety and comforts of "decent" middle-class life and its "civilized" ways. Here in the New
Orleans underground, they don't have to pretend to be normal or decent.
They carry out deviant lives, moving from bursts of creativity to binges of
drugs, sex, and alcohol. And this freedom to shape one's biography and
develop new rules of urban living creates fascinating lifeworlds beyond yawn-
ing mainstream life.

But to know these lives, we must first learn about the places they trans-
gress.

TRANSGRESSIVE LIVES IN THE CITY OF TRANSGRESSION

It's not easy to carve out a creative life of transgression on the urban fringes.
Living on the precarious edge of the late-modern city is fraught with dangers.
Like real estate, it's all about location, location, location.

But when it comes to living on the margins, New Orleans is an ideal location for creative urban dwellers to develop their transgressive biographies. The city's Vieux Carré and its surrounding historic neighborhoods offer them cultural habitats in which to survive, even prosper, in their endeavors. The five main characteristics of transgressive urban habitats help us understand how New Orleans culture is a mecca for the urban creative misfits of late modernity.

First, transgressive urban habitats must offer opportunities to find free food and cheap drinks throughout the week, especially at local bars and restaurants. Second, these habitats must have a tourist-intensive economy that draws millions of people annually, numbers that bring a large influx of tourist dollars into the local economy, including to the nontraditional economy of the urban outsiders. Third, the city's mainstream culture must endorse, or at least generally accept, the presence of a vibrant, transgressive, and creative poor class. Cities rich in history and steeped in cultural traditions celebrate, even if only symbolically, the traditions of their longtime communities, especially those associated with struggle and resilience. Urban areas with large populations living in close proximity, often with rich and poor living in the same neighborhoods, foster unique relationships between the privileged and subordinate classes. Although such relationships between the powerful and powerless can lead to sometimes violent confrontations, it can also breed a sense of familiarity and closeness between two otherwise polarized and highly differentiated classes. This is especially true in New Orleans, where the demographic profile in many neighborhoods changes from street to street. In such cases, living in poverty does not carry as much of the social and cultural stigma that it does in other places. Fourth, transgressive urban habitats need to have neighborhoods with ample foot traffic consisting of tourists who liberally give money to panhandlers, buskers, and hustlers as well as to formal and informal street merchants. Fifth, these habitats must encourage the development of subcultures of urban tribes that can emerge in spaces close to the surrounding tourist-intensive pleasure and entertainment zones. These makeshift urban tribes develop loose networks of support that make life in the cultural underbelly of the city possible.

These five crucial elements exist in New Orleans and make the city an attractive enclave for urban fringe dwellers.

Culture of Lagniappe

It's easy to get free shit everywhere in the Crescent City. Free dinner is served up almost nightly at a wide variety of the city's watering holes, so long as you drink—which isn't really a problem for most New Orleanians—and as long as you pay for those drinks. Lagniappe is all about the respect that business owners pay to customers for frequenting their establishments. New Orleanians might not always be polite, but being polite is not the same as having manners, and the culture of lagniappe exemplifies good manners. The people of New Orleans might be weird, but they have their own peculiar brand of etiquette, passed down from the old aristocracy. For those with a working knowledge of those manners, it becomes easy to use them to survive on the streets, hustle money and find drugs, eat and sleep, drink and smoke, pursue romance and creative endeavors, find sex and satisfaction, and survive. The urban fringe dwellers of New Orleans learn where to eat for free nearly every night of the week while avoiding the stigma associated with welfare cards, soup kitchens, and food pantries—though these are viable alternatives as well. Though many New Orleans restaurants "crackerjack"[6] their prices for tourists, gentrifying yuppies, and the city's old-money elites, cheap and good food is available if one knows where to look.

As I write this, Mick's Irish Pub, a local watering hole in the Mid-City neighborhood, offers free red beans and rice for dinner every Monday—an early-week tradition throughout the city—and everything from hot dogs to chicken pot pie on other weekdays. Pal's Lounge, another neighborhood bar in Mid-City, also offers free red beans and rice on Mondays, hot wings on Wednesdays, and pasta on Sundays. Just pay three or four bucks for an Abita beer and tuck in to your free, hot meal. No one will notice you are down and out, particularly if you're with friends.

For just $1.25, you can hop on a streetcar that will take you from the French Quarter to places on St. Charles Avenue that shuck twenty-five-cent and fifty-cent raw oysters for smart locals and savvy tourists every weekday. In Bucktown, just over the Orleans parish line, one can get fresh boiled crawfish for under $3 a pound. The Hare Krishna Temple on Esplanade Avenue serves free vegetarian Indian food every Sunday evening. In the Uptown neighborhood, the Saint has a happy hour that offers some of the cheapest booze in town, with cans of Pabst Blue Ribbon for $1.50, two-for-one well drinks, and a dollar off everything else, along with a free jukebox. Scores of other bars and restaurants all over the city offer great deals, enabling people to eat and drink on little to no money every day of the week.

A few bucks in your pocket also provide access to some of the best nocturnal entertainment in the country. At Marigny's Siberia or Decatur Street's Spit Fire, you can catch free burlesque shows throughout the week. Although many of the Frenchmen Street drinking and music venues, such as d.b.a. and Blue Nile, charge a cover, if you arrive an hour before the music ends, it's free. From Hot Eight Brass Band to Eric Lindell to the Tremé Brass Band, you can shake your ass to some of the best music around. And free music vendors still exist all over the city as well.

Tourist-Intensive Economy

The tourist economy of New Orleans brings in close to ten million people annually, numbers nearly equivalent to its pre-Katrina days, and more than $6.81 billion a year.[7] *Condé Nast Traveler* recently ranked New Orleans as one of the "best big cities" in the United States, one of the "top 25 cities in the world," and as the number-one "U.S. City for Business Travelers."[8] Sauntering through the pleasure and entertainment zones in and around the French Quarter, especially during the city's many spring and fall festivals, one sees a city inundated with travelers from around the world. Droves of tourists in khaki shorts, white sneakers, and knee-high socks walk the streets, "hand grenade" cocktails in hand, as clowns and jugglers, sword swallowers and opera singers, gutter punks and beggars, street performers and vendors compete to titillate them.

Who gets these billions of tourist dollars? While it is clear that the economic and political elites benefit from the billions pouring into the city annually, the vast majority of New Orleanians, especially those most responsible for producing the culture, work low-wage jobs. Although tourism employs more people than any other economic sector in the city, it serves up the lowest annual salary. While about one in four of the jobs in the city, close to 27 percent, are related to the tourist industry, most of these jobs pay poverty or near-poverty wages.[9] Despite the increasing profits, which disproportionately benefit the city's elites, today's poverty rates are similar to what they were before Hurricane Katrina changed so many other things about the city in 2005.[10] According to the U.S. Census Bureau, nearly 30 percent of New Orleans residents live below the poverty level, and poverty disproportionately affects black residents, at almost 37 percent below poverty. Still worse, and most troubling, about half of all black children live below the poverty line.[11] The tourist industry might provide billions to corporate powerbrokers, but precious little of that wealth trickles down to the thousands of rank-and-file hospitality-sector workers, who face scant prospects of any significant upward professional mobility. It does, however, provide an ample resource for those who exist on the economic fringes of the city and within its cultural underbelly.

The tourist industry also provides opportunities for the creative poor to make money, from the sale of art and poetry, jewelry and soap, and food and music. The heavy foot traffic of tourists with money to spend creates sustainable opportunities to beg, hustle, and busk. Many buskers earn incomes higher than the average wages of many of the city's service-industry employees. Gutter punks with clever signs and sleepy dogs make enough for beer, Pop-Tarts, and some junk for a high. Hustlers rely on the adage that a sucker is born every minute, using well-worn short cons to extract hard-earned money from tourists.

To find food, places to sleep, alcohol, drugs, and companionship, many of these hustlers, buskers, and beggars form informal support networks. Some of them exploit the system, supplementing their incomes by making use of social safety nets such as welfare, disability, and unemployment insurance. "Home is where I lay my head," one busking squatter says laughing. His friend adds, "We prey on people's kindness."

Virtuous Poor

Most New Orleanians live on the lower end of the socioeconomic spectrum and well below the national average. While U.S. households earned a median income of $53,657 in 2014, New Orleans residents earned a median income of $36,964. The city's overall poverty rate of 28 percent is almost double the official national poverty rate of 14.8 percent.[12] New Orleans is a culture that knows what it means to be poor, and, at least for the majority of the people in New Orleans, living as a part of the working class and in near-poverty conditions is not so harshly stigmatized.[13] As dichotomies such as "first world versus third world" dissipate in late modernity, we begin to see first-world conditions in the third world and third-world conditions in the first. New Orleans offers a potent example of this, especially in the neighborhoods where the rich and poor live in juxtaposition. The monster of poverty is no strange beast here. The bumper-sticker slogan "Third World and Proud of It" applies to the living conditions of New Orleans, especially inside historically black neighborhoods and the many mixed working-class neighborhoods such as Gentilly and New Orleans East. For many people living in New Orleans, poor is the norm, and the multi-billion-dollar tourist industry benefits from this. Still, for those willing, low-paying jobs in the tourist industry are abundant and fairly easy to obtain. Many restaurants and bars will offer some type of job to anyone willing to work, if only for a single day.

It has been noted that the United States is a rich country filled with poor people who mock themselves and glorify the rich.[14] In contrast, New Orleans culture is built from the resiliency of people finding creative solutions to their collectively experienced structural problems. As a result, most people live poor and value people based on qualities other than the size of a bank account. The majority of New Orleanians, I would argue, hold no loathing for the poor. Most of us are poor or damn near it.

In New York City, the saying goes, small talk often begins with "What do you do for a living?" In New Orleans it's all about how you *do* your living. It's more about how you tell stories than your social status, occupational status, or location within the structures of power and ideology. Rather, a sense of belonging is achieved—and social status earned—through one's lifestyle choices and storytelling capabilities, through creative pursuits and genuine connections made with people. We value Bourdieu's cultural capital over economic capital—and it's a distinctly New Orleans cultural capital.[15]

What I'm describing here is not liberal sympathy for the poor. Rather, it's simply that it's normal to be poor, or close to it. We have jobs that pay poorly—in many cases, very poorly.[16] But they pay, and somehow we make it—even if it keeps many of us at the fringes. Some restaurants on Decatur Street offer one-day-only employment washing dishes, for the person who arrives first in the morning. Café du Monde, once a major employer for women in the Vietnamese community, now employs many white Americans from middle-class stock, both locals and transplants. It may not be so easy in the Big Easy, but in New Orleans, as Sly Stone philosophized, you can make it, even if only barely, if you try.

High-Density Foot Traffic

As Jane Jacobs asserted, a healthy sidewalk serves as a central mechanism of a city's vitality and safety.[17] The relatively compact area of the French Quarter and its surrounding historic neighborhoods (Faubourg Marigny and Bywater, the Seventh and Ninth Wards, and Tremé) brings into close proximity the lives of a wide variety of urban inhabitants. The many subcultures and eccentrics running around highly concentrated areas of the city form urban tribes. These tribes offer locals and transplants access to a loosely structured division of labor and to the social network necessary to survive. Many of these tribes share their resources and knowledge, from knowing the neighborhoods with the best squats to combining their financial resources, from busking in the streets to sharing affordable housing. They also provide networks of emotional support, gathering to meet in the city's many coffeehouses and bars, conducting "song shares," and dining communally in the private spaces of their homes and various neighborhood speakeasies disguised as art studios.

Despite the many music venues now requiring hefty cover charges, free music abounds in various neighborhoods throughout the city. Mobs of people bar hop between music venues, absorbing the jazz and funk spilling out into the streets. These crowded streets offer young urban hustlers ample opportunities for a ping in the tip jar and produce an abundance of potential customers for the informal street merchants selling their crafts and talents. On Decatur Street near Molly's at the Market and Coop's Place, for example, struggling artists and gutter punks, with their dogs at their feet, play guitar, fiddles, and flutes; sing melodies; beg and panhandle; and display items of all sorts for sale, from photographs to poems. As Decatur nears Canal Street, black youths from the nearby Seventh Ward tap dance for tourists. Frenchmen Street in the Marigny hosts some of the greatest music in the city—brass bands at the Blue Nile, New Orleans–style jazz at d.b.a., world music at Marigny Brasserie, 1920s jazz at the Spotted Cat, contemporary and traditional jazz at Snug Harbor, sounds from a variety of traditions at Yuki Izakaya, and local New Orleans artists at Three Muses, among many others.

Outside the nearby venues and shops, people make money by selling paintings or writing love letters on typewriters, all while nearly naked women twirl batons and shake their asses, young brass bands toot their horns, and guitarists coax tunes from their battered instruments. While most subsidize their incomes through informal street vending and busking, some also work as part-time bartenders, waiters and waitresses, teachers' assistants, baristas, dishwashers, busboys, food deliverers, cashiers, strippers, oyster shuckers, ghost-tour guides, community organizers, volunteers, and other types of workers in the many tourist venues of the city.

Subcultures of Urban Tribes

Subcultures of urban tribes develop and emerge from the concentrated tourist spaces of entertainment and pleasure-seeker zones and their surrounding neighborhoods, where those who make a living in the French Quarter often live. Lacking access to institutional resources, those who exist in the cultural underbelly of New Orleans develop their own subcultures, or urban tribes, that find solutions to the conditions of living down and out in the city. This is reminiscent of Duneier's findings that show how magazine vendors create informal networks of support to keep their work environment sustainable— from getting to work on time and avoiding criminal activity and excessive drug use to protecting their street materials and conforming to some sense of mainstream norms.[18] Similarly, those living on the edges of New Orleans society form creative informal networks of support that make possible a life-style that rejects conventional and mainstream norms and routine, brain-draining jobs of disenchantment.

The members of these urban tribes, which are usually centered on creative activities, from music to art to writing, pool their financial resources, busking and begging together while sharing rent in cheap accommodations. They sometimes lend money to one another and network with others in the tribes, connecting people within and among tribes. These urban tribes roam the city—attending poetry readings, pop-up restaurants, formal and informal music venues, art studios doubling as unlicensed bars, song shares and informal music jams, house gatherings, comedy shows, local late-night television shows, burlesque shows, theater events, community talks, support groups, twenty-four-hour bars and restaurants, and a variety of other community activities.

THE HARD AND SOFT CITY IN NEW ORLEANS

The Rabanesque view of the city juxtaposes the hard and soft city, in this case, the hard city of New Orleans and the soft city within these liminal "hard" spaces.[19] The hard urban grid of the French Quarter, with its Spanish-style architecture and the city's river-carved crescent shape, provides the space for the cultural production of the soft city, where its inhabitants impose their subjective will on the city. Only the French Quarter forms a true grid; the other neighborhoods for the most part bend and wind, conforming to the will of the mighty Mississippi River. The next section takes a closer look at the French Quarter neighborhood before widening in scope to some of the others in the city's collection of oddly shaped neighborhoods.[20]

Vieux Carré

The Vieux Carré, or French Quarter, is bordered by North Rampart Street to the north, Esplanade Avenue to the east, the Mississippi River to the south, and Canal Street to the west. The oldest neighborhood in the city, it forms a grid thirteen blocks long and six blocks wide. Although the culture of New Orleans is produced in the adjacent faubourgs, that culture is sold in the French Quarter. The old Vieux Carré boasts colonial Spanish architecture and is the site of tourist attractions and local dives—the center of pleasure and entertainment. Here one can find Café du Monde, the Cabildo, Bourbon Street's glitzy and sleazy twenty-four-hour drinking and music establishments, Royal Street's art studios and antique shops, the venerable Jackson Square, the iconic St. Louis Cathedral, the Moon Walk promenade along the Mississippi, the French Market, and the old U.S. Mint. The French Quarter is a place of balconies and breasts, cocktails called hurricanes and hand grenades, beignets and cafe au lait, voodoo and T-shirt shops, strip clubs and gay bars, riverboats and casinos, restaurants and music venues, hustlers and the hustled, antique shops and art studios. It is the heart of the city—but the heart depends on its arteries.

Strip clubs, or "Sallie Mae" establishments, cluster on Bourbon Street from Toulouse to Iberville.[21] Young, nearly naked women often stand outside chatting with bouncers and attracting the attention of tourists strolling up and down Bourbon Street. Nearby, young black men posing as shoe shiners hustle unsuspecting tourists with the locally known "Twenty dollahs I can tell ya where you got dem shoes" line (for the right price, they can even tell you what street you got them on). Meanwhile, winos and druggies walk the lively Vieux Carré streets drinking forty-ounces and Mad Dog 20/20 (sometimes called crack wine, bum wine, or street wine). The subcultures of "deviants" and outsiders create much of what the French Quarter is known for, attracting both tourists and a variety of new types of settlers to the city.

Tourists travel to New York to find out what the locals are up to; in New Orleans, locals go to the French Quarter to see what the tourists are up to. Here in New Orleans, we have a contradictory relationship with tourists. Most consume in the French Quarter a watered-down version of the local culture, far removed from the neighborhoods that produce it. Some litter the streets with their "go" cups, vomit and piss where they please, and listen to Bourbon Street music, which is in most cases a far cry from the city's real music scene. Others treat those working in the service industry like servants. Still others—especially the obnoxious and the drunk—are disrespectful, condescending, and stiff their servers of our well-earned tips. Many tourists walk around the French Quarter with little appreciation for the rich culture nearby. While we view these visitors with suspicion and a degree of contempt, we also realize their vital contribution to the economy and thus to our livelihoods.

The neighborhood's small residential population experienced a 9 percent decrease since Hurricane Katrina, dropping from 4,176 in 2000 to 3,813 by 2010.[22] French Quarter residents are generally white and wealthy—that demographic composed about 89 percent of the population before the storm. This remained basically unchanged: the white population was 87.6 percent of the total after the storm in 2010. The black population remained below 5 percent both before and after the storm. The average rent in the French Quarter sits around $1,111 to $1,200 a month. The already high amount of vacant housing increased over 5 percent, from 37.4 percent before Hurricane Katrina to

43 percent after the storm. Perhaps the most substantial change is the sharp increase in the annual household income of French Quarter residents, which rose from $80,518 to an estimated $123,254 in post-Katrina 2014. Meanwhile, the percentage of people living in poverty slightly increased from 10.8 percent to 11.7 percent in 2014.

Faubourg Marigny

Adjacent to the French Quarter on its eastern edge, "the Marigny," as most locals know it, is bordered by St. Claude Avenue to the north, Franklin Avenue to the east, the Mississippi River to the south and Esplanade Avenue to the west. This area, once known as a distinctly local spot, was one of the first places to gentrify in New Orleans, with a population consisting largely of gay workers in the French Quarter's service industry. The Marigny is shaped like a drunken triangle and boasts a blend of residential shotgun houses and colorful nineteenth-century Creole cottages as well as an entertainment district with some of the best music in New Orleans, especially on Frenchmen Street and near the intersection of St. Claude and Elysian Fields. Sometimes referred to as the "new" Bourbon Street, Frenchmen Street (described further in chapter 7) between Decatur and Royal Streets attracts droves of tourists, gutter punks, and street entertainers—from the Frenchmen Street poets to brass-band musicians to dozens of food and craft vendors—and they set the social and cultural scene for some of the main characters discussed in later chapters.

A thriving and relatively new nocturnal urban scene near the corner of St. Claude and Elysian Fields flourishes as both locals and tourists pack music and drinking venues such as the largely heavy-metal scene Siberia, the naughty karaoke bar Cajun Pub, and the mixed-music venue Hi-Ho Lounge, which plays everything from Deejay Soul Sister to live brass bands, as well as the sexy eclectic theater and bar Always Lounge. It's party central in the Marigny Triangle, with dozens of local watering holes, such as the R-Bar, Dragon's Den, Buffa's Lounge, and Checkpoint Charlie's, as well as rough and edgy local drinking establishments like the John, Lost Love Lounge, Mimi's in the Marigny, St. Roch Tavern, and the Phoenix. On most nights, the entertainment district of the Marigny becomes a playground for both locals and tourists to drink and dance the night away.

Faubourg Marigny has experienced a slight 5.5 percent decrease in population since Hurricane Katrina, dropping from 3,145 to 2,973 people.[23] The number of black people in the neighborhood has declined from almost 18 percent to just below 13 percent since the storm; the percentage of whites increased from 72.6 percent to about 75.5 percent. Meanwhile, the average costs of rent rose from its pre-Katrina days of $876 to $981 a month. The Hispanic (of any race) population also slightly increased, from 6 percent to 6.8 percent of Marigny's overall population. The amount of vacant housing increased about 4 percent, rising from 16.6 percent before Hurricane Katrina to 20.3 percent after the storm. The average household income increased over $20,000 from its pre-Katrina days of $49,166 in 2010 to $69,796 in 2014. Meanwhile, the percentage of people living in poverty slightly decreased, from 24.1 percent to 21.4 percent.

Bywater

The Bywater, which is situated "by the waters" of the Industrial Canal and Mississippi River, serves as the hipster mecca of the city's artistic and music scene. Bywater sits below the Faubourg Marigny neighborhood and is bordered by St. Claude Avenue to the north, the Industrial Canal to the east, the Mississippi River to the south and Franklin Avenue to the west. It was once a place of grit, a strange and hidden neighborhood filled with eccentric locals and old shotgun houses.

The neighborhood has seen a 35 percent decrease in its population since Hurricane Katrina, dropping from 5,096 to 3,337 people.[24] The more significant demographic shift occurred in the decline of the black population, from about 61 percent to 33 percent of the overall population; the percentage of whites increased from 32.4 percent to about 56 percent, a clear indication of gentrification. Meanwhile, the average rent rose from a pre-Katrina $833 to $985 a month. The Hispanic (of any race) population also slightly increased, from 4.8 percent to 6.7 percent of Bywater's overall population. Unsurprisingly, the amount of vacant housing increased from 17 percent before Hurricane Katrina to 29.4 percent after the storm. Another clear sign of gentrification, the average household income dramatically changed from its pre-Katrina days of $37,455 in 2000 to an estimated $57,219 in 2014.

Meanwhile, the percentage of people living in poverty decreased from 38.6 percent to 18.5 percent.

This strange and quirky neighborhood has rapidly gentrified in recent years, perhaps more than any other in the city. One sign posted in front of a house on Alvar Street close to Dauphine reads:

> There are 355 AirBnBs in Bywater tax free the cost of this was devastating my ssi 7.00 2014 tax 4.00 2015 2,200 for 2016 2500 I live in poverty and fear while seeing others get wealthy air bnb your killing me.

A bartender named Bobby, who has served happy-hour drink deals and lunch specials at BJs Bar on the corner of Lesseps and Burgundy streets since 1990, tells me the place has been around for about eighty years. Up until the 1980s, it was a beer parlor with living quarters for a family in the back. When I asked him about the urban underbelly of New Orleans, he said, "What urban underbelly? It's nothing but hipsters and yipsters and yupsters and all the other -sters. I can't afford to live here. This place has priced me out." Bobby wonders where he is going to live next.

The Bywater is one of the great neighborhoods of the city, and that adds to its attractiveness for both hipsters and the economically superior bourgeois bohemians. Although primarily a residential neighborhood boasting some of the city's oldest and most colorful shotgun houses and Creole cottages, numerous mom-and-pop businesses pepper the area. The bar Bud Rips on the corner of Burgundy and Piety opened in 1960. The bartender tells me that the building housing Bud Rips has been a bar since 1865. It was previously called Tacs, then Bersierons, before being changed to Bud Rips in 1965. Many locals in the neighborhood believe it is the third-oldest bar in New Orleans and the second-oldest continuously open bar. (Lafitte's Blacksmith Shop in the French Quarter is believed to be the oldest, and Bourbon Street's Old Absinthe House the second oldest). When asked if any of the locals still frequent the establishment, the bartender comments that the hipsters and the yuppies are now the locals. Some old heads still straggle in, though they are few and far between nowadays.

Vaughan's Restaurant on the corner of Dauphine and Lesseps sells poboys, daily specials, and seafood—as well as a healthy supply of beer. It looks worn down and dilapidated. Beneath a completely rusted roof, the façade of the building is covered with chipping paint. It's definitely been through its fair share of storms. Ceiling fans hang beneath a tin canopy along the sidewalk outside, where tree stumps serve as seats near tables alongside a long, forest-green bench. Neon signs reading "Abita Beer," "Mardi Gras," and "Miller Lite" hang in the windows, along with a sign apparently from the owner that says "No Loitering, No Crack Selling, No Cat Selling, The Facts— Mr. Samuel. NOPD will be called" and another pseudoadvertisement listing "Wildlife Specials." It serves food only when live music is also on the menu, which is usually on Thursdays. Tourists will frequent the bar on music nights, mainly because it was featured in the HBO television show *Treme*. The bartender, Sarah, says the bar has been around since the 1800s. It's had the same owners since 1981. Sarah says it is the only bar that has not changed in the Bywater. She has been here since 1997, lives in the neighborhood, and says that after Katrina, residential rent prices skyrocketed. She knows of places that once ran around $450 to $500 a month that are now renting for $1,200 a month. With a sigh she says, "It seems like more young professionals are in the neighborhood that don't drink, at least in our bars. As a result, we have less customers that frequent the place." On Thursday nights, when live music plays, entry comes with a ten-dollar cover charge—but there is also free red beans and rice. And the famous New Orleans jazz trumpeter Kermit Ruffins still plays here sometimes.

Jimmy, of Jimmy's Grocery on 4201 Dauphine, has owned the business for twenty-one years and complains that his property value jumped from its pre-Katrina $5 to its current $30 per square foot. He feels that the city elites have allowed a gentrification process to occur, complete with brand-new lofts catering mostly to a rich white population that has displaced much of the local community. Economic times are tough for these small Bywater mom-and-pop stores.

Bywater is crawling with bohemians, musicians, artists, writers, punks, goths, and just about anyone who lives an alternative lifestyle. But apart from the hipster invasion, Bywater is still home to a working-class mix of black and white residents. Many of the old establishments still serve cheap and affordable food and beer—such as Frady's One Stop Food Store[25] on Dauphine Street, twenty-four-hour small bodegas selling cheap fried catfish and booze, and late-night bars like J&J's Sports Lounge on the corner of

Dauphine and France Streets. Coffee shops cater to both the blue-collar locals and the invading hipsters. Live music and dance venues such as BJs Lounge and the gritty Saturn Bar also attract large local and transplant crowds. It's common for both tourists and newcomers to the city to meet a Bywater resident out in the French Quarter or Marigny neighborhoods and take a wild adventure into the magical Bywater night. It would not be the least bit surprising if a night in the Bywater involved an array of booze, sex, magic mushrooms, and pot, all with a pinch of existentialism and collective transcendence.

Faubourg Tremé

Abutting the French Quarter to the northwest is the newly famous Tremé neighborhood, bordered by Broadway to the north, Esplanade Avenue to the south, North Rampart Street to the south, and St. Louis Street to the east. It's one of the oldest and historically most prosperous black communities in the United States. The HBO show made it famous, but to locals, it has always served as one of the hearts and souls of the city. Tremé birthed jazz and brass bands; it's where the culture of New Orleans was created and where it's still made today. Its historic shotgun and Creole cottage homes sit in the shadows of the French Quarter and the oak trees of Esplanade Avenue's neutral ground. Tremé is believed to be the first area in the United States where black people could buy land,[26] and it's famous for Congo Square, where jazz was born from the drumbeats of African slaves. It's a culturally rich mecca of New Orleans and vital to our history. It's a lively neighborhood where one can feel the thickness of the culture that birthed the roots of this city.

The neighborhood has experienced about a 47 percent decrease in its population since Hurricane Katrina, plunging from 8,853 to 4,155 people.[27] Overall, white people are moving into Tremé as the black population declines. This significant demographic shift, another indication of gentrification, resulted in the decline of the black population from 92.4 percent in 2000 to 74.5 percent in 2010; the percentage of white people increased from 4.9 to 17.4. The average costs of rent in the neighborhood today ranges from about $674 to $858 a month. The Hispanic (of any race) population also slightly increased, from 1.5 percent to 5.4 percent of Tremé's overall population. Unsurprisingly, the amount of vacant housing increased from 19.4 percent before Hurricane Katrina to 37 percent after the storm. Although information about the average household income for the post-Katrina years is missing ($26,895 in 2000), the percentage of people living in poverty from 2000 to 2014 decreased from 59.6 percent to an estimated 44.3 percent.

Central Business District (CBD)

On the western edge of the French Quarter sits the present-day CBD. Today, it is a bustling downtown, complete with gleaming office buildings, rushing taxis, and armies of professionals in suits. Once upon a time, however, it was known as "the Swamp" and was one of the most notorious underworlds of the Mississippi River towns, known for its hedonistic debauchery, feared by the police and all middle-class locals with any common sense, with its "mazes of narrow streets and alleys teeming with gamblers, murderers, footpads, burglars, arsonists, pickpockets, prostitutes and pimps, and ruffians who would gouge out a man's eye or chew off his nose for the price of a drink."[28] The Swamp attracted "half-savage Kaintuck Keelboat crews . . . gamblers, politicians and harlots who fleeced them. . . . Runaway slaves [hidden] in sheds and tents far back in the trees . . . [where] gaggles of snarly-haired prostitutes hunted giving themselves to forty men a night . . . [where] the city's poor squatted in squalid cabins."[29] All the wild "mads" and ruffians engaged in the deepest of vice and sins in the "incredible jumble of cheap dance halls, brothels, saloons and gaming rooms, cockfighting pits, and rooming houses. . . . [Where] women were notoriously 'abandoned to lewdness' . . . [and] tough 'ladies of the evening' [attracted] hundreds of flat boatmen."

Today the CBD is bordered by Claiborne Avenue to the north, Canal Street to the east, the Mississippi River to the south, and Calliope/I-10/Earhart Expressway to the west. It is now filled with politicians and the bureaucrat-infested City Hall; now a new type of "snarly-haired prostitute" roams its streets.

The CBD has an even smaller residential population than the French Quarter, with 1,794 people before the storm, rising to 2,276 people in 2010, a 21 percent increase.[30] Whites composed the majority of the racial composition of the neighborhood both before and after the storm, increasing from 55.2 percent to 62 percent in 2010. The black population decreased from 32.9 percent to 23.1 percent over that same period. The average cost of rent in the

CBD is similar to the French Quarter, at somewhere between $1,100 to $1,200 a month. Like many other New Orleans neighborhoods, the already high amount of vacant housing increased from 21.5 percent before Hurricane Katrina to 29.3 percent after the storm. As for French Quarter residents, the average annual household income towers above the average for Orleans Parish (in 2000, before the storm, $59,354, up to $62,880 in 2014), slightly increasing from a whopping $92,976 to $94,878 in 2014. Meanwhile, the percentage of people living in poverty remained high after the storm, increasing slightly from 32.3 percent to 33.6 percent in 2014.

Mid-City

Mid-City is bordered by Metairie Road and City Park Avenue to the north, Orleans and St. Louis to the east, Claiborne to the south, and Earhart Boulevard and Monticello Avenue and Bamboo Road to the west. It's a large, mostly middle-class neighborhood about five or ten minutes from almost anywhere in the city. The streetcar running along Canal and Carrollton streets in Mid-City makes the French Quarter and City Park easily accessible to anyone with $1.25 in their pocket. The neighborhood sits between Lake Pontchartrain and the Mississippi River, the long Bayou St. John connecting the two waters. The 1,300-acre City Park, bigger than Manhattan's Central Park, lies in the heart of Mid-City, with its centuries-old live oak trees and hanging Spanish moss.

The residential population of this large neighborhood has decreased about 26.5 percent, from 19,909 in 2000 before Hurricane Katrina to 14,633 in 2010 after the storm.[31] The black population significantly dropped from its pre-Katrina days of 64.3 percent in 2000 to 55 percent in 2010; the white population increased during this same period, from 23.2 percent to 27.3 percent. The Hispanic population (of any race) increased from 10 percent to 15.2 percent, attributable largely to demands for cheap immigrant labor following Hurricane Katrina. Today, Mid-City residents pay between an estimated $787 and $953 a month for rent but, judging from personal experience, sometimes much more. The amount of vacant housing nearly doubled from 13.3 percent before Hurricane Katrina to 25.7 percent after the storm. The average annual household income remained relatively consistent before and after the storm, changing slightly from $43,224 to $44,026 in 2014. Meanwhile, the percentage of people living in poverty remained about the same, changing slightly from 32.1 percent to 33 percent in 2014.

Mid-City is the place where many older transplants end up settling, should they decide to make New Orleans home. It's a neighborhood of locals with many dives and watering holes—such as Pal's Lounge, Mick's on Bienville, Banks Street Bar and Grill, Finn McCool's, Twelve Mile Limit, Parkview Tavern, and Bayou Beer Garden—as well as restaurants catering to locals, including the "unofficial Jazz Fest headquarters" Liuzza's; the romantic Café Dega; the cheap Juan's Flying Burrito; the old-school Italian restaurant Venezia; the politician- and cop-patronized Betsy's Pancake House; Angelo Brocato's Italian Ice Cream Parlor, serving up the best ice cream and coffee in town; the late-night hamburger place Bud's Broiler; the bourgeoisie Crescent City Steakhouse, serving the city's elite and corrupt; the neighborhood Mediterranean diners Fellini's Café and Mona's; the classic New Orleans restaurant Mandina's, with its famous trout almandine and turtle soup; the local writer's haunt Fair Grinds Coffee House; and the well-known Chinese restaurant Five Happiness. We know food in New Orleans, and Mid-City boasts some of the best of it. Although the neighborhood is not as trendy as Bywater, it's a local community where New Orleanians return after a night of play in the city's wilder and sexier neighborhoods. Mid-City also includes the beautiful tree-lined areas of Esplanade Avenue and its wide neutral ground, along which locals spend hours walking and talking.

Lower Ninth Ward

This former cypress swamp produced some of the greatest New Orleans musicians, including Antoine "Fats" Domino and Kermit Ruffins.[32] The Lower Ninth Ward, the "Lower Nine," is bordered by the Industrial Canal to the west, the Southern Railway railroad and Florida Avenue Canal to the north, the parish line to the east, and St. Claude Avenue to the south. Holy Cross is an adjacent neighborhood also considered part of the Lower Nine, bordered to the west and east by the levees of the Industrial Canal and Jackson Barracks near the St. Bernard Parish line and to the north and south by St. Claude Avenue and the Mississippi River. This neighborhood was almost completely inundated with water during Hurricane Katrina. Today it's a mostly black working-class neighborhood that is beginning to show the first signs of gentrification, with hipsters and middle-class whites moving into the area.

The locals dub the Industrial Canal Levee separating the Ninth Ward from Bywater the "End of the World," a bit of local poetry to capture the feeling one gets standing near the edge of New Orleans on a large levee overlooking the canal emptying into the great anaconda-shaped Mississippi River. This neighborhood, because of its isolation from the rest of the city, problems with inadequate drainage systems, and vulnerability to flooding, took a long time to develop. But it also made the area a cheap place to live, attracting poor blacks and European immigrant laborers. Members of the black New Orleans community organized in the 1870s to help Lower Ninth Ward residents, many of them former slaves, with their concerns, especially with the drainage and pumping system. Although drainage systems were installed in 1910 and 1920, the 1923 Industrial Canal that connected Lake Pontchartrain to the Mississippi River further isolated the neighborhood from the rest of the city. The neighborhood still seems distant from the rest of New Orleans, retaining an almost rural feel so many years after its late development.

The development process continued into the 1950s as the growth of retail shops and small corner stores along St. Claude Avenue expanded deeper into the Ninth Ward. As commercial activity increased, a second bridge was built—officially named the Judge William Seeber Bridge but locally called the Claiborne Avenue Bridge—spanning the Industrial Canal at Claiborne Avenue and connecting the city to the Lower Ninth Ward.

Adversity often creates resilience, and the Lower Ninth Ward serves as a fine example. City neglect fostered a culture of activism in the community. Consequently, the area birthed one of the milestones of the civil rights movement: the McDonogh Three—the nickname given to three six-year-old black students at the Ninth Ward's McDonogh No. 19 Elementary School— who in 1960 became the subject of one of the first challenges to racial segregation after it had been made illegal in *Brown v. Board of Education*.

Today, the Lower Nine is a quiet residential neighborhood that retains the grit of its past. Much of the neighborhood suffers from third-world conditions, with its poor, pothole-marred streets and stock of abandoned and vacant housing, businesses, and schools, especially the old and colorful graffiti-filled buildings on St. Claude Avenue. The primarily residential neighborhood has a few commercial properties and small businesses—from a visitors' center and a café, a used-clothing store and a variety of churches, a daycare and small market—and a variety of closed-down and vacant businesses: empty cleaners and gas stations, a vacant restaurant called Eatin' at Holmes, and a muffler shop with a permanent "On Vacation" sign. But many of the neighborhood's residents are not on vacation and struggle while facing increasingly insurmountable obstacles to rebuild their community and the houses and businesses within them.

The numbers reveal that Hurricane Katrina has drastically changed both neighborhoods. The Lower Ninth Ward and Holy Cross experienced catastrophic damage from the storm, some of the most severe in the entire city. The total population of both neighborhoods combined dropped about 71.5 percent, from 19,515 people in pre-Katrina 2000 to 5,556 in post-Katrina 2010.[33] In recent times, both areas have been home to a predominantly black population, and they remain so after the storm. The Lower Ninth Ward's black population of 98.3 percent in 2000 slightly declined to 95.5 percent in 2010; Holy Cross's black population slightly increased during this same time, from 87.3 percent to 89.3 percent. Meanwhile, the white population of the former increased from .5 percent to 1.8 percent during this same period; the latter neighborhood's white population decreased from 9.4 percent to 6.9 percent. One of the most striking shifts in the neighborhood pre- and post-Katrina is the extreme spike in vacant housing, from 13.9 percent in the Lower Ninth Ward (15.3 percent in Holy Cross) to 48 percent in 2010 (41.1 percent in Holy Cross). Today, the average rents are somewhere between $500 and $1,000 per month. In both neighborhoods, the average annual household income dropped, from $37,803 to $33,328 in the Lower Ninth Ward and from $44,269 to $42,219 in Holy Cross. While the percentage living in poverty slightly decreased in the Lower Ninth Ward, from 36.4 percent to a still statistically high 33 percent, the percentage of people living in poverty increased in Holy Cross from 29.4 percent to 38.9 percent.

Seventh Ward

The Seventh Ward is a neighborhood of odd angles and even odder characters.[34] It is bordered by A. P. Tureaud Avenue, Agriculture, Allen, Industry,

St. Anthony, Duels, Frenchmen, and Hope streets to the north, Elysian Fields Avenue to the east, St. Claude and St. Bernard avenues, North Rampart Street, and Esplanade Avenue to the south, and North Broad Street to the west. *Les gens de couleur libres*[35]—or "free people of color"—once made up the bulk of the population of this thriving and formerly prosperous community just east of Faubourg Tremé on the opposite side of Esplanade Avenue. Many of these free people of color were light skinned, French-speaking Creoles. They had more access to economic opportunity thanks to their lighter skin tones. These educated and successful Creoles created a culturally and economically vibrant community. In the 1850s, many of them—equipped with education, skills, and business acumen—opened up family businesses, funeral homes, barbershops, and other community establishments, leading to economic prosperity. All this changed in the aftermath of the Civil War. The Jim Crow laws and the one-eighth "octoroon" rule lumped anyone darker than a brown paper bag into the category of Negro. While the Jim Crow laws segregated blacks and whites into unequal social groups, they also inadvertently led to the development of jazz, by bringing Creoles and black Americans together. As Creoles and former slaves shared for the first time the same segregated community spaces, they played their distinct forms of music together, blending the sounds of the European-influenced Creoles with the traditional music of newly freed African slaves.

The late 1960s proved disastrous for the Seventh Ward after real-estate developers constructed the new elevated I-10 interstate loop right down the center of Claiborne Avenue, often referred to as the "black people's Canal Street,"[36] a place where successful black-owned and Creole businesses had until then thrived. This business district, along with all the live-oak trees on the wide neutral ground, were cleared to make way for the new stretch of interstate, largely destroying the community's culture and economic vitality. People left, homes were abandoned, and businesses failed, and the city did nothing to ameliorate the dire conditions. The Seventh Ward is now home to mostly poor black residents who have developed a resilient culture to confront the challenges the community now faces. The political and economic elites of the city are suggesting tearing down the Claiborne corridor to pave the way for new urban development. Community members are understandably suspicious of the city elites' intent, but young white hipsters taking advantage of the cheaper rents are rapidly moving into the area.

The current demographics of the neighborhood looks more or less the same as pre-Katrina New Orleans, albeit with an increase of whites moving into the neighborhood.[37] It is still a predominantly black neighborhood, 87.4 percent, down about 6 percent from its pre-Katrina levels; the white population has doubled since the storm, 6.6 percent and counting. Vacant housing significantly increased, from 16.2 percent pre-Katrina to 38.3 percent in 2010. The average household income has decreased from $36,544 in 2010 to a current estimated household income of $29,840. Poverty has also increased since the hurricane, as it has in other mostly black neighborhoods, reaching an estimated 41.9 percent, an almost 4 percent increase in poverty.

Roaming through the area, one encounters rows of abandoned and boarded-up shotgun and double-shotgun houses lining the potholed streets. Mixed in with them are others that appear well cared for. None of them have lawns, but just about all of them have New Orleans–style stoops, where residents often sit, whiling away the afternoons. Many of the streets resemble the crumbling, pothole-filled streets of third-world cities like New Orleans' sister, Port-au-Prince, Haiti.

Parts of the Seventh Ward are witnessing a resurgence, especially on St. Bernard Avenue, with the emergence of venues that attract locals and transplants—like the art house and underground-music venue United Bakery Gallery, the extremely friendly and welcoming local bar Poor Boys across the street, and nearby Sidney's Saloon, where locals and transplants dance to local New Orleans band King James and the Special Men.

St. Roch

This neighborhood, like the Seventh Ward, boasts a proud history of having once been home to free black people before the Civil War.[38] It was once called Faubourg Franklin and birthed great jazz music and musicians, including the ragtime pianist Jelly Roll Morton. Developed in 1830 with the creation of the Pontchartrain Railroad, which connected the Faubourg Marigny with Milneburg, the St. Roch neighborhood grew in the twentieth century. In the 1920s, St. Roch was known as a laid-back and racially diverse residential section of the city. A large baseball field also attracted fans of the sport. The St. Roch Market, an affordable public fish market, was once a staple for the community. It was renovated in the 1940s and sold to a private company that operated it as a fish market and purveyor of po-boys and gumbo.

In the late 1960s, the I-10 Interstate Loop cut through St. Roch, disrupting the community and leading to neighborhood abandonment and decay. The city did little, if anything, to prevent the neighborhood's decline. The effect of the I-10 interstate is quite visible in St. Roch; the neighborhood on one side of the highway became a less desirable place to live. Tenants left; houses fell into disrepair. Despite the disruption, the people worked to keep their neighborhood intact with playgrounds, parks, green spaces, the historic St. Roch Chapel and Cemetery, and other places that facilitate community activities.

The demographic portrait of St. Roch looks much like the nearby Seventh Ward. The neighborhood is still composed of a mostly black population (86.8 percent), about 5 percent lower than its 2010 pre-Katrina population. Meanwhile, as in the Seventh Ward, the white population nearly doubled to 7 percent in the years following the storm. As expected, vacant houses in the neighborhood increased post-Katrina from 16.6 percent to 37.7 percent; both average household income and poverty increased in the wake of the storm. It is estimated that average household incomes decreased by nearly $10,000 after the storm, falling from $38,970 to $29,136; poverty increased slightly more than 3 percent, to 40.5 percent. The newly renovated St. Roch Market has caused controversy within the local community: Its crackerjacked prices cater to yuppies while excluding many of the local black working-class community residents.

Uptown

The Uptown section of New Orleans encompasses various neighborhoods, including Carrolton Avenue, the Garden District, the Irish Channel, Central City, and the Lower Garden District. Although the boundaries of Uptown are debatable, even among New Orleanians, the area makes up about one-third of the city, from Monticello Avenue to the north, to Earhart Boulevard to the east, to the Mississippi River to the south and west. Although most of the scenes in this book take place in and around the French Quarter and its neighborhoods, Uptown—or the huge section of the city west of Canal Street—is an important part of New Orleans. Uptown is as diverse as the city itself—from poor neighborhoods to the old moneyed mansions of the streetcar-lined St. Charles Avenue to cheap happy-hour oysters to the swanky 1883 Columns Hotel to centuries-old cemeteries to posh Tulane University, which most New Orleanians can barely even afford to wave at while rolling by on the streetcar. Uptown has some of the best snowball vendors, such as Hansen's, as well as some of the best po-boy shops, such as Guy's on Magazine Street and Parasol's in the Irish Channel neighborhood. Coffee shops line Magazine Street, and music venues such as the famous Tipitina's romantically hug the Mississippi River on Tchoupitoulas Street. Uptown is also home to Mardi Gras Indian tribes in the rough-and-tumble Central City neighborhood.

While the Audubon area of Uptown consists of a large (around 15,000 people before and after the storm) white (around 85 percent before and after the storm) and wealthy ($149,793 before and $158,590 after the storm) population, with high rents above city averages and lower percentages of vacant housing, much of this large neighborhood is composed of quite different demographics.[39] The Uptown statistical area, between Jefferson and Napoleon avenues running north to south and La Salle and Magazine Avenue (near the Mississippi River), is composed of 5,984 residents, about 700 people below its pre-Katrina numbers. The black population has decreased from 36 percent pre-Katrina 2000 to 21.8 percent in 2010; the white population increased from 57.8 percent to 69.9 percent. The Hispanic population grew slightly about from 3.5 percent in 2000 to 5.4 percent in 2010. Vacant housing increased from its 2000 pre-Katrina level of 10.2 percent to 15.9 percent in 2010. Average household incomes significantly increased from 2000 to 2014, from an already relatively high annual $75,996 to $124,456; the poverty rate dropped from 23.9 percent to 11.6 percent.

To show Uptown's varying demographic shifts from one part of the neighborhood to the next, the Freret area immediately east of the Uptown statistical area described above has a smaller population of 1,715 people (2,446 before the storm), composed mostly of black residents who now make up about 72 percent (as opposed to 82.7 percent before the storm) of the neighborhood's population. Meanwhile, the white population grew about 5 percent after the storm to about 18.7 percent of the overall population. Hurricane Katrina hit this neighborhood hard, increasing the amount of vacant housing from 16.9 percent in pre-Katrina 2000 to 30.6 percent in 2010. While the average household income increased from $55,931 to $60,563, the amount of poverty remained the same at about 34 percent of the population.

GUTTER PUNKS, HIPSTERS, AND BOURGEOISIE BOHEMIANS

The human geographer Richard Campanella of Tulane University identifies three types of groups settling into the older neighborhoods of New Orleans: gutter punks, hipsters, and bourgeoisie bohemians.[40] These groups are moving in droves into the historic Tremé, Bywater, Seventh Ward, Ninth Ward, and Marigny neighborhoods, among others, and they are gentrifying and transforming the human geographies and local cultures they find there.

Gutter punks are anticonformist youths from various classes who live by panhandling, begging, dumpster diving, and squatting in the vacant houses of New Orleans and sometimes in the city streets themselves. Decatur Street in the French Quarter and Frenchmen Street in the Marigny Triangle are the quintessential gutter-punk spaces, with heavy foot traffic, cheap beer and food, lively music, police tolerance for panhandling and art vending, sympathetic tourists, hole-in-the-wall bars, semiprofessional hip locals, and a general celebration of gritty beatnik and bohemian culture. In the heart of New Orleans's Decatur Street, near drinking establishments such as Molly's on the Market, the Abbey, Aunt Tika's, Coops Place, the Hideout, and Turtle Bay (just to name a few), gutter punks spend hours panhandling, playing music, writing songs and poetry, smoking cigarettes and joints, and talking about life. Gutter punks survive on the streets by panhandling and begging, often with somewhat clever signs, like "got a dime for some slime." Many of these gutter punks hail from middle-class families and have bank accounts they can access if times get too tough. Every so often, a homeless gutter punk can be found sneaking off to a "magic window"—an ATM.[41]

The open and hostile rejection of mainstream society—its puritan values, moral contradictions, boring and mentally destructive jobs, self-importance, and commodity fetishism—is at the center of the gutter punks' creation of a subculture of miscreant youths (and sometimes the not-so-youthful), developing their own alternatives, however destructive, to living a mainstream life.

While the "fuck-authority-and-establishment" mentality is a valiant creative form of resistance to structural marginality and institutional submissiveness, it also leads to individual self-destruction, and members of this subculture experience long periods of unemployment, alcohol and chemical use and sometimes outright abuse, and extremely unhealthy living conditions. They are godawful dirty and smell like hell, too. To the French Quarter business owner, they are aggressive nuisances that threaten business. To hipsters, they are emblems of the urban experience, though the two groups don't always get along so well. While many gutter punks see hipsters as middle-class inauthentic snobs pretending to live on the edges, hipsters see gutter punks as lazy, mendicant, unproductive members of society without ambition.

The term "hipster" describes urban dwellers that teeter on the edge of mainstream society and sanitized gutter-punk life.[42] Hipsters outwardly reject mainstream conventions while unevenly navigating that same mainstream world and quasi-subcultural belonging. Urban hipsters celebrate superficial diversity, tolerance, and difference while forming a loosely knit homogenized group of formally and informally educated people, mostly from middle-class stock. Some work part-time jobs mostly in the service industry and sometimes in education, community activism, and cultural organizations. These part-time jobs allow hipsters to remain somewhat dedicated to lifestyles that reject conventional jobs and middle-class structure while also providing the opportunity to pursue their various artistic endeavors.

Bourgeoisie bohemians are the middle- and upper-middle-class professionals—often from areas such as New York City, San Francisco, and Los Angeles—who move into the once-affordable historic neighborhoods of the city in search of "authentic" urban life but without the annoyances of America's larger cities. While most of the city's neighborhoods recovered 90 percent of their pre-Katrina population—and some have gained population since the storm—the demographics of New Orleans looks very different from its pre-Katrina figures, especially when considering race. Although New Orleans lost around 11,500 white residents, the city lost nearly 100,000 members of its black population.[43] While still a majority-black population, the city has become whiter, and many of the new whites hail from outside the metropolitan area. These new transplants largely consist of post-Katrina volunteers, artists, poets, teachers, performers, musicians, and educated big-city dwellers.

In effect, as property values rise, gentrification and the influx of new transplants to these historic neighborhoods push out many locals, many of whom
have been residents of these communities for generations. This transformation has far-reaching consequences for New Orleans's culture.

———————

The people of New Orleans internalize the culture of the city in the same ges-
ture as they project their own unique characters back onto the city. They
simultaneously shape the city as the city shapes them. In turn, the people of
New Orleans, *both* locals and transplants, place their subjective imprints
onto the city as they scratch out their biographies, imposing their wills upon
the objective culture of New Orleans. Traveling the urban spaces of New
Orleans with the city's transplants and locals reveals a fascinating world
where various subcultures emerge to develop their own unique and creative
solutions to the structural problems posed in late modernity, all within the
context of a postindustrial urban metropolis. But to live fully in this world,
one must become immersed in the down-and-out life of the New Orleans
cultural underbelly. This involves roaming with these creative city dwellers
as they ply their unique trades, dance in music venues, drink in watering
holes, smoke pot in their rented Creole cottages, sell their artistic wares on
sidewalks, stroll along the Marigny, and sit in coffee shops discussing their
idealistic, world-changing ideas, shaping New Orleans, creating new ideas in
art and literature, and waxing poetic about the city and its many characters.

CHAPTER 3

LIVING DOWN AND OUT IN NEW ORLEANS

A *plongeur* is a slave, and a wasted slave, doing stupid and largely unnecessary work. He is kept at work, ultimately, because of a vague feeling that he would be dangerous if he had leisure. And educated people, who should be on his side, acquiesce in the process, because they know nothing about him and consequently are afraid of him.

—George Orwell, *Down and Out in Paris and London*

BARTENDING AT CAFÉ BEIGNET ON BOURBON STREET

It's Memorial Day weekend at Café Beignet on Bourbon Street, right on stripper row in the Vieux Carré, about three blocks from Canal Street. It's a time of year when many travelers from around the world dip down to New Orleans before the oppressive summer heat takes hold of the "northernmost Caribbean city"—we're only thirty degrees above the equator. At 9 a.m., with the smell of Cajun breakfast filling the air, the old French- and Spanish-style courtyard called "Musical Legends Park" offers an inviting space for tourists as they begin their day amid the statutes of some of the city's music heroes. Steamboat Willie and Friends prepare to play jazz, Dixieland, and ragtime for their first set, much to the delight of eager tourists thirsty for New Orleans tunes and specialty cocktails.

As the line for beignets and Cajun eggs and hash browns builds, I open the wooden shutters that almost completely hide the enclosed courtyard bar, preparing it for what promises to be another busy day of getting tourists drunk. Many things must be done before the bar opens: The ice compartment needs filling, the day's fruit garnishes prepared, the daiquiri machines and hurricane container filled, the napkins and plastic cups stocked, the thousand dollars of register money counted, and so on. Perhaps the most important thing is to check where everything is placed, especially the most essential bartending tools and drink ingredients. Since each bartender organizes the bar differently, placing well drinks, lemon mix, sweet-and-sour mix, simple syrup, shakers, bitters, fruits, and juices in their own preferred and idiosyncratic locations, it's tricky to find everything. It's a treasure hunt, and treating it as such keeps the experience from becoming an otherwise aggravating endeavor.

Prior to doing any preparation, opening the shutters is the first fateful step toward opening the bar and facing the morning's first tourist. It's typically a fat and thirsty soul with a big, loony smile. In a loud and unmodulated voice, and with a red-as-a-boiled-crawfish face, he asks rhetorically, "You're open right? I'll take four bloody marys." "I'm on it," I reply. After ducking and diving through throngs of tourists to retrieve the ice, more thirsty tourists queue up. First, I prepare the four bloody marys: Place four cups on the bar, fill each with ice, add two shakes of celery salt, two dashes of Worcestershire sauce, the house-made "bloody mary magic juice," and the house-made bloody mary mix, complete with spicy tomato juice and well vodka. The concoction is topped off with two olives and two pickled string beans. "That's thirty-two dollars please," I say—and before the money is in my hand, more tourists arrive, while still others shout their orders. They order the house irish coffee and a drink we call café de chile orchata, which consists of rum, half-and-half, cinnamon, vanilla extract, and coffee with chicory. The orders keep coming, from mardi gras mornings (Mardi Gras–brand vodka, Kaluha, and cream) to tequila sunrises. There's no longer time to stock the bar for the shift.

Three more tourists order drinks. Beer, that's easy, as is the next order of a ten-dollar Southern Comfort mango daiquiri. Most domestic beers sell for six bucks; imports sell for seven, though some of the local beers also sell for seven. Holy hell, somebody actually orders the cinnamon apple cocktail, made with Fireball whiskey, Apple Pucker, and pineapple juice. The famous New Orleans hurricane, easy to pour and sweeter than sugar, sells for nine dollars. Now there's a request for a sazerac, but where the hell did the Angostura and Peychaud bitters go? The sweat drips from my face in the already steamy and humid day, and tourists crowd around, waiting for their turns to order. I can feel the impatience building. But where did the bitters go? After thirty seconds of frantic searching, I find it. The crowd gets thirstier. Shake, shake that Sazerac and circle that lemon peel over the rim of the plastic cup. "Mojito? Coming up," I say, immediately wondering in what absurd place the mint leaves might be hiding. Inside the refrigerator, perhaps? No dice. Certainly not on the bar counter. Drawers? No. I suddenly remember a walk-in cooler outside the bar. Mint leaves! Running back into the bar, it's mint-leaf muddling time. Oh, shit, where's the fucking muddler? Someone left the damn thing in the cooler between the whipped cream and the lemons. Muddle, muddle, and shake, shake does a fine mojito make. Grab the six bucks and press "2" on the register four times to ring in the order. Shove in the money and turn around only to hear, "Bombay and tonic please, with a lime."

Six boisterous ex–frat guys approach the bar; the women they are with sit in the courtyard waiting for them to return with drinks. They order a margarita rocks, a bloody mary, a double Jack and coke with a lime, a pinot grigio, a Bud Light, two Abita Ambers, a seven and seven, and a Tanqueray and tonic. I begin pouring the beers, which gives me time to devise a strategy for making the rest of the drinks—but only if I can remember the orders. I can't. I make one at a time and have to ask them to repeat their drink orders twice while taking what feels like forever to make them.

Another customer asks for two beers. The manager asks, "You got everything?" I reply, "Of course, got this," clearly lying to both the manager and myself. The Bud Light pours smoothly into the pint-sized plastic cup from the tap—until it suddenly runs dry, sputtering and coughing explosively, drenching my shirt. The keg needs changing. Just as I turn around, I see that a red light on the strawberry daiquiri machine has lit up, indicating the machine is running low. I change the keg, pour two beers, and hand them to the customer. There's a brief break in the orders as the musicians take a break. Time to prepare for the next onslaught.

The bar needs more of everything: napkins, fruit, plastic and Styrofoam cups, simple syrup, well drinks, and bar towels, all stocked and prepared for use. Leaving the bar requires pushing past tourists in the courtyard and into the kitchen, where cooks and workers frantically prepare orders. The sound of tough men cursing reverberates throughout the kitchen, the air of which carries the blended smell of New Orleans beignets and jambalaya. Run, stop, and sidestep the sweaty short-line cook to make way, jump over a misplaced box, squeeze between the stairway and another cook to jump over steps leading to the towels and plastic cups. Outside, it's the smell of coffee and chicory, and the powdered sugar that tops the beignets takes flight as tourists commit the rookie mistake of exhaling while biting down. The courtyard is sticky sweet and foggy in the hot humid air.

At the bar, a new line of customers waits for me to take their orders. After pouring some beers, it's time to make the strawberry daiquiri, a process that involves mixing seventy-two ounces of premade strawberry mix and thirty-six ounces of well rum. This goes into the machine along with two big containers of Mississippi River–sourced tap water. The strawberry mix is loaded with high fructose corn syrup. It's diabetes in a cup. The few moments between customers allow for retapping the Bud Light.

The Steamboat Willie Jazz Ensemble—composed of clarinet, trombone, tuba, string bass, piano, and drums, along with East St. Louis–born Larry "Steamboat Willie" Stoops leading on trumpet—prepares for another set in the now-quiet courtyard, where the only sounds are of tourists mumbling through their mouthfuls of beignets. The courtyard faces directly onto Bourbon Street, where passersby can peek through a black gate with a corridor that allows entry into the venue.

Suddenly Steamboat Willie's trumpet gives forth a resounding call, like Leviticus's horn proclaiming jubilee throughout the land. As the music fills the humid air, the tourists stop dead in their tracks, tilt their heads to the side, and begin an awkward and uncertain shuffle toward the source of the music. While the musicians play, seduced tourists gravitate to the siren call.

New Orleans is full of sirens. Just outside Café Beignet's courtyard, nearly nude strippers prance around, attracting tourists and businessmen like the creatures of Greek myth, luring their prey to their lair from the ocean of Bourbon Street.

Behind the bar, my merciful lull ends. Tourists en masse take their positions, barking out orders.

Some tourists, for reasons unclear—maybe because of my olive skin—speak to me in something vaguely resembling Spanish. Some folks—especially western Europeans—refuse to tip. Others give three or four dollars for every order. Some ask stupid questions: "What do I want?" "What's good?" Others ask for directions to places, advice on where to eat or drink, places to stay. "Can I take these drinks to go?" Of course you can. This is New Orleans.

Once the onslaught ends, I take a deep breath, then contemplate the horrible spectacle before me: Sticky simple syrup coats the bar's surface, mix bottles are out of place, chunks of fruit litter the bar, bottles of fruit juice as well as half-and-half—of course all capless—are scattered around, shakers and muddlers are displaced, spills of every variety cover nearly every surface of the bar, empty bottles are strewn about, and dirty and empty cans and containers wait to be dispatched to the already full garbage bin.

While cleaning up, two smokers ask for ashtrays—and then miss them with each cigarette flick. A drunk man of fifty harasses the musicians with silly questions, making them signal to me to cut him off. A woman flirts, asking for an extra shot in her mimosa. A Japanese man accidently drops a thousand yen—worth about nine bucks American—and gives it to me when I retrieve it for him.

Eight hours after it started, my shift ends at 5 p.m. The bar needs to be prepared for the poor bastard who will relieve me. This involves restocking everything and cleaning just about everything, from the drainage holes on the floor to the entire bar area, including the bar tools and utensils. I count my tips: $150 dollars, plus $6 dollars an hour—but I owe the register thirty dollars for a missing credit-card receipt.

After the prolonged assault, a sense of shock, a drained buzz. After pouring so many drinks for so many drunken people, one wonders if my decision to self-medicate with alcohol is the best idea. But the alternative to drinking is simply to stare into the void. After two or three beers, I feel better—not relaxed or comfortable in any way, just better somehow.

Like most French Quarter workers, I commute by bicycle—the other option being the inefficient public-transportation system—since parking in the quaint but cramped neighborhood is next to impossible. It takes about thirty minutes to ride my bike home to Mid-City late in the evening. My home consists of an air mattress in someone else's kitchen. Between the fatigue from work and the alcohol I drank afterward, I'm left with a numb sort of unfeeling. But it eases the shitwork exhaustion.

Not a single critical, creative, or intellectual thought passed through my brain today. There was no time for it. But now I wonder: How did we as a society get to the point where selling one's mind and body in exchange for a wage seemed like a good idea? Why do we hold on to the ideological notion that "hard work" is masculine, as if a "real man" works hard or a "real woman" finds liberation through the world of work?

THE TOURIST AND SERVICE INDUSTRIES AS CHATTEL SLAVERY

"You get used to it." That's what they typically say. The other bartenders explain to me, as a new bartender on the scene, about the world of work in the tourist industry. Other workers in the service industry echo these same sentiments. But who wants to get used to working forty to sixty hours a week, the same tasks, day in and day out, at least five days a week? Regular work destroys the creativity and drive of the vast majority of human beings. Perhaps this is the tragedy of modern life for so many Americans and for the vast majority of people around the world, who must sacrifice huge chunks of their life for low wages that offer just enough to keep them alive so that they can continue working their jobs. As the German-born American poet and novelist Charles Bukowski wrote in his 1975 novel *Factotum*: "How in the hell could a man enjoy being awakened at 8:30 a.m. by an alarm clock, leap out of bed, dress, force-feed, shit, piss, brush teeth and hair, and fight traffic to get to a place where essentially you made lots of money for somebody else and were asked to be grateful for the opportunity to do so?"[1]

The vast majority of Americans, essentially, are bullied into going to work every day, in places that they do not want to go, in order to engage in activities they do not want to do. Our entire society is arranged around this idea. Sometimes, the oppressed develop a sort of Stockholm syndrome: They come to identify with their masters and work willingly in their interests, even if it goes against their personal interests. This allows the same cheerful robot mentality to continue.

In talking about how people internalize oppression, Chomsky explains:

At one time in the U.S., in the mid-19th century, working for wage labor was considered not very different from chattel slavery. That was the slogan of the Republican Party, the banner under which northern workers went to fight in the Civil War. We're against chattel slavery and wage slavery. . . . Free people do not rent themselves to others. Maybe you're forced to do it temporarily, but that's only on the way to becoming a free person. . . . You become a free man when you're not compelled to take orders from others. . . . That's an Enlightenment ideal. Incidentally, this was not coming from European radicalism. There were workers in Lowell, Massachusetts . . . saying this around that time. It took a long time to drive into people's heads the idea that it is legitimate to rent yourself. Now that's unfortunately pretty much accepted. So that's internalizing oppression. Anyone who thinks it's legitimate to be a wage laborer is internalizing oppression in a way which would have seemed intolerable to people in the mills 150 years ago.[2]

The women of Lowell Mills in Massachusetts lacked university educations and had little to no exposure to the radical thinking of the Marxists or the progressive sociologists of the time. But it does not take a Ph.D. or a Marxist or a sociologist to realize that wage work closely resembles slavery, and the Lowell women, perhaps more aware of this than the most educated of us, perhaps especially the most educated, understood exactly the ramifications of wage work. These women sang a song[3] during the strike of 1836, fighting against industrial exploitation and oppression:

Oh! isn't it a pity, such a pretty girl as I—
Should be sent to the factory to pine away and die?
Oh! I cannot be a slave,
I will not be a slave,
For I'm so fond of liberty
That I cannot be a slave.

If you hate freedom for the people, love the system of wage labor. If you love freedom for the people, attack the oppressive forces that shackle us to coercive labor. The women of Lowell knew it. The people of Haiti know about slavery and wage labor as well, and most of them lack formal education. They have a proverb that says, "If hard work was so good, the rich would have stolen it long ago." Perhaps it is wise to listen to the Haitians and to all those who have a history of dealing with slavery—from the Lowell women to the exploited working people of the world to the indigenous populations we continue to devastate.

ARRIVING DOWN AND OUT TO NEW ORLEANS

My ten-week experiment in living down and out in New Orleans began on May 18, with only a few possessions on my person—a pair of black leather shoes and a pair of sneakers, a pocketbook calendar, digital audio recorder, money clip, a decoy "stick-up" wallet, toiletry bag, pen, laptop computer in its bag, notebook, pocket knife, three changes of clothes, cellphone and chargers, and five days of undergarments—along with $100 in my pocket. I'd already paid my first month's rent of $400 for a room in Mid-City, though it turned out not to be a room at all but an air mattress in a drab and dark dining area of an old rundown apartment, with absolutely no privacy. It also had no air conditioner—lovely for the hot summer months. I borrowed $100 from my "Boris,"[4] my childhood friend Bobby, to purchase a bike from a horrid chain retail store (that shall remain nameless) of the type that has consumed the American landscape and all but eliminated the mom-and-pop retail shop. With new transportation, a pot to piss in—albeit no window to throw it out of—and a hundred smacks in my pocket, it was time to find gainful employment.

I had a one-page résumé listing my contact information; childhood home on Odin Street[5] as my residence (today it's an empty lot); educational level as a bachelor's degree in sociology from the University of New Orleans; employment history of mainly service-industry jobs[6] in New Orleans prior to moving to New York for graduate school; my real and manufactured skills—bartending, busing and waiting tables, cooking, writing, world travel, and public speaking); and my "insights," stated as: "I'm a native son of New Orleans who, after graduating with a sociology degree at UNO, spent the last few years traveling the world and making fascinating discoveries. I'm now returning to my city." My employment history simply provided the names of restaurants without any contact information or dates employed. Many of these restaurants no longer existed. My references included an urban studies professor at UNO; my friend Lightfoot, who works as a medical repairman; and Bobby, a fireman and plumber. In subsequent interviews, no one ever verified my past employment history, checked any references, asked about my education, or cared about my traveling experiences or why I'd been gone for so many years. I could have easily made the whole thing up. Perhaps just having a résumé, speaking the dominant mainstream English dialect, and not being black put me at an advantage over others, at least with some prospective employers. Job applicants who give the impression of being middle class hold a definite advantage over the working classes, especially black residents, who often get the short end of the stick.

I walked around the French Quarter for two days looking for employment in one of the many hospitality and service-industry jobs of the French Quarter, filling out applications with dozens of bars and restaurants, from dingy holes in the wall to upscale establishments catering to wealthy tourists. It was the typically slow summer season, so not even the Lucky Dog vendor company was hiring. But employment possibilities remained. I received three or four calls for interviews, including Larry Flynt's Barely Legal Hustler Club; the bar and restaurant 13 Monaghan, on Frenchmen Street; and Café Beignet, in the 300 block of Bourbon Street. I interviewed at Larry Flynt's and was told to call again Friday. 13 Monaghan offered me an interview but later called to give me the bad news. Café Beignet offered me an interview and a probationary shift to prove myself. Although my limited experience bartending in New Orleans happened over ten years ago, I assured them that I would get the job done. I eventually filled out a W-2 form, a Homeland Security form, and a direct deposit form. (Everyone gets paid via direct deposit.) I also signed off on a policy dictating some bullshit rules regulating employee behavior—including cellphone use, texting, drinking—as well as other expectations about work duties and customer interactions. It's only six bucks an hour with tips, but at least a man can still get a city gig without much hassle.

Although my experiment working down-and-out jobs in New Orleans included various types of employment—cleaning rooms, pantomiming, writing poetry on commission for tourists, performing as a clown in a sideshow freak show as a glass stomper—it all began at Café Beignet.

BARTENDING ON BOURBON STREET

Café Beignet is located on 311 Bourbon Street. It is home to the "Musical Legends Park," a courtyard with statues of various famous New Orleans legends, including Pete Fountain, Fats Domino, Allen Toussaint, Irma Thomas, Chris Owens, Al "Jumbo" Hirt, Ronnie Kole, and Louis Prima. Café Beignet's outside bar sits in the courtyard overlooking the statues where the musicians perform.

Music plays all day and night at Café Beignet. The management hires musicians for full-time gigs, like Steamboat Willy and a drummer who claims to have played for Dr. John and helped write the song "Right Place, Wrong Time." (That story could not be verified, and it's probably worth noting that Dr. John has sole songwriting credits on that classic slice of New Orleans funk.) One section of Café Beignet serves coffee, tea, and food, mostly a watered-down version of standard New Orleans fare, as well as sandwiches, po-boys, and fried seafood. As the name of the place implies, they also serve beignets, which are just fine. Customers order food and coffee at one section of Café Beignet, while at another section a full bar caters to a boisterous tourist crowd. Bands play all day in the courtyard, with occasional intermissions, and bartenders, baristas, and short-line cooks keep the food and drinks pouring nonstop.

Café Beignet stays open between 9 a.m. and midnight on weekdays but closes at 1 a.m. on weekends. There are a total of six bartenders, including one of the managers, Chris. There's always a manager on duty, as the city government requires for all drinking establishments, though Chris is the only manager that knows how to bartend. He's a professional bartender full of clever quips and insights as well as knowledge about the world around him. Employees get one free meal per shift, unless it's seafood, in which case it's half off.

At first I worked the day shift but after a couple of weeks moved to the night shift, which started at 5 p.m. and went through to 11 or 12 at night, though we often stayed later cleaning the bar and courtyard while waiting for the manager to count the money in the drawer and announce our credit-card tips for the day. Tourists make up the overwhelming majority of the customer base, from regional tourists from the Gulf South to others from France, England, and Japan. The band plays mainly New Orleans blues, jazz, and R&B, starting soon after the first bartending shift of the day begins.

In general, once the music plays, people start drinking. They sit in the courtyard drinking coffee and/or alcohol, listening to music and talking away the humid afternoon. My first drink order served there was two Blue Moons, two Cokes, a Heineken, a margarita, and two Abita beers. I made my first sazerac soon after and heard the young woman tell her male companion after taking her first sip, "I *told* you to order this!"

For the rest of that first day, I found myself mixing margaritas, mint juleps, sazeracs, and mojitos, along with gin and tonics and tequila sunrises. While I was admittedly relearning how to mix drinks from my previous limited experience over ten years earlier, I made about $70 on tips (some cash, others on credit cards), plus $48 for eight hours of work at six bucks per hour. The credit-card tips get placed in our weekly checks, subject to taxation, and sent to direct deposit in the weekly check delivered every Friday. We also must claim a percentage of money in tips based on some percentage of the overall money the bar earned each shift. I grew increasingly confident in my bartending abilities after every shift, including on my second day, when I completed ninety orders (not ninety drinks but ninety *orders*, which can contain many drinks each) and earned about $100 (split between cash and credit card) in tips. The managers appreciated my efforts and put me on the night shift. As it turns out, they put me on the night shift to accommodate the needs of a new employee. Also as it turns out, somewhat counterintuitively, the night shift typically earns a bartender considerably less money. That's because the day shifts have only one bartender and no cocktail waitresses. Night shifts must split money equally between two bartenders and two cocktail waitresses. The night shift is also considerably more difficult, since two bartenders must make drinks for patrons bellying up to the bar as well as for the packed courtyard being served by the cocktail waitresses.

One day I was short on the register and was forced to pay it back directly from my tips. (The register starts at $1,000; after the shift, the manager counts the money of the night's sale in the back manager's office.) Chris explained that register was $25 short, stating, "It happens to the best of us." I paid the difference without complaint. Chris explained that the most common reasons for a short register at the end of the night include failing to collect signed credit-card receipts, pressing a button on the register too many times for an order, and returning too much change. Either way, bartenders must pay the difference out of their own pocket. Money was tight, and losing twenty-five bucks affected my plans for beers that night with Lightfoot and, worse, for buying food the next day.

Bartenders and cocktail waitresses rely on trust to split the tips. Though bartenders manage all the money served from drinks, they must trust the cocktail waitresses to place all of their tips into the collective tip jar, and the cocktail waitresses must trust that the bartenders will report all of their tips the same. Of course, it is hard to know if the bartenders or the cocktail waitresses are pocketing some of their tips, which amounts to theft. On one occasion, one of the bartenders decided that we would withhold two hours of tips from a cocktail waitress who we decided was being lazy on the job—taking walks around the French Quarter, smoking cigarettes, doing anything but taking orders for drinks. We also sometimes rang up thousands of dollars in drink orders in a single night only to discover that we each had only made a total of $240 in tips, or $60 each—significantly less than what one would expect for such a large volume of business. Although we suspected a particular cocktail waitress was pocketing her tips, there was no way to prove it. Cocktail waitresses can easily not report their tips, especially since most are collected in the courtyard, away from the eyes of the other employees. Eventually, she was fired, officially because of "customer complaints," but we all knew it was really because we thought she was cheating the rest of us.

My bartending colleague Moira has been serving drinks for thirty-five years, mainly in the French Quarter. Moira is a tough New Orleans woman who has made her life as a bartender and manager in the hospitality industry. She's blonde, stout, in her fifties, and claims to have started bartending when she was seventeen years old. Moira does things her own way. She begins the shift by taking charge of the place, giving various orders to the other bartender (me) and the cocktail waitresses. She claims to makes drinks "old school," saying, "I ain't no mixologist, darling. I'm a bartender. My momma was a bartender and her momma was a bartender and her momma." For example, her mint juleps and mojitos are simple: muddle the mint in a plastic cup, add ice, rum, simple syrup, and voila. She says this is Bourbon Street, not some fancy bar. And if she is going to serve a hundred or two drinks per shift, things need to be made simple. Moira is highly organized and likes everything clean; she even wants all the bills in the register facing the same direction. The first day she worked with me, she left her shift like a bat out of hell. Splitting shifts, she says, between four people (two bartenders and two cocktail waitresses) on a slow night does not make sense. I also wanted to leave, but she would have none of that.

The cocktail waitresses also serve as bar backs, filling ice compartments, taking out the garbage, putting away empty bottles, and so on. Though some cocktail waitresses treated me well, others seemed to despise me.

Working with service-industry employees gives one a keen sense of their focal concerns, and their sense of morality centers on these concerns. At the Café Beignet, workers quibbled excessively over tips and on making sure no one gets over on them. For example, Moira refuses to give customers free tap water.[7] She says that if she is going to work, they are going to pay. Giving customers tap water for nothing equates to customers "getting over" on her, taking advantage of her labor. Instead, she charges them for a small bottle of water, three dollars to quench the thirst of a hot New Orleans day. That this money does not go to her but to the bar owners that exploit her labor makes little difference. The focal concern is to prevent people from getting over on her.

Other focal concerns are centered on respect and morality. This is especially the case for nontippers and disrespectful customers. Moira tells a story about a business owner who was a customer where she was tending bar. The owner treated her rudely when ordering a drink. She told him off, saying, "I'm not lower than you just because I'm a bartender. I work just like you do. Do you treat your customers the way you just treated me?" He later gave her a twenty-dollar bill, stating, "I'm sorry if I offended you in any way, shape, or form." "You know what?" Moira said. "He will probably never treat another service worker again that way."

Conversations at work often involved obsessive conversations about tipping, a reasonable topic of discussion given that most bartenders and other service-industry workers depend on tips to make a living. In fact, many service-industry jobs in New Orleans pay minimum wage ($7.25 as of this writing) or worse, leaving many if not most workers in the tourist industry dependent on tips for daily survival. It's unsurprising that the employees might get angry when tourists stiff them, and it happens all the time. Some workers occasionally reprimand customers for not tipping. Moira once told a customer, "Really, you're not going to include my service? This tip jar is for tips, sir." The man gave a "Well, I never" look and walked away. Another cocktail waitress, in a fury, marched the tip jar straight to customers sitting at a table.

We produced grand theoretical postulates about which European and Asian cultures tipped and which types of Americans tipped poorly.[8] Tipping was an issue particularly with tourists from countries without a history of tipping and ignorant of American culture. Many such tourists simply don't realize that Americans in the service industry depend on tips for their livelihood. Tips make the difference between eating or starving and paying for rent. At worst, some tourists remain willfully ignorant and refuse to tip their servers regardless of the cost and disrespect it shows to the worker.

One day, a Japanese businessman ordered drinks for all ten people in his party—two mint juleps, two sazeracs, a margarita, a ramos gin fizz, a gin and tonic, a cinnamon applesauce, an Abita, and a strawberry daiquiri, a total of $67 dollars. While about a dozen other customers waited to order, I placed all the glasses on the bar, muddled the mint, put all the ingredients into the various shakers, poured the beers, filled the daiquiri cup, and added the final touches and fruit garnishes. After watching me make all of these drinks quickly—but well—the man handed me a hundred-dollar bill. I made change at the register, making sure, as any smart bartender does, to provide the small bills necessary to allow the customer to provide a tip. The man took out his wallet and stuffed every bill into it.

At first, I found the obsession with tips funny. But after working for a couple of weeks, the humor subsided, and instead I found myself having unspeakable thoughts about the nature and character of those who don't tip. What kind of human wouldn't pay for services rendered? Who would allow another human to serve them and give nothing in return? Making someone labor and refusing to pay for services rendered go beyond the disrespect of simply robbing someone of hard-earned cash. But we found clever ways to procure tips, to secure our hard-earned money. Moira would always tell tourists what they owed "without service"—that is, "That will be $25, without service." Sometimes it would work. One time I simply explained to a tourist that people in the United States who work in the service industry depend on tips to make a living. We always kept on both sides of the bar a huge tip jar with a rubber toy shark propped up eating a dollar bill to encourage the customers to tip. People seemed to feed the tip jar more when they saw a bunch of money already in it. Sometimes the cocktail waitresses, out of sheer frustration, would stand next to the table they just served, waiting for their tip.

There were other annoyances. Sometimes people puked all over the place. (Who do you think cleans it up?) Others stumble around as drunk as a sailor on leave. People talk to you incessantly about the most unimportant details of their dull and dreary lives. A superior decides your work schedule every week, which takes away a lot of control over your life. Some people hit on you; others argue with you about the amount of booze you've poured in their drink.

Despite the many aggravations and annoyances of working in the tourist economy on Bourbon Street, there are some advantages to bartending, include cheap drinks on service-industry night. Further, after working in the politically correct multicultural neoliberal academic environment of the university, it was refreshing to speak like a genuine bona fide human again. We call each other "darling" and "love," speak openly about what is on our minds, yell at one another during disagreements and forget about it the next day—all brothers and sisters again. We make jokes about the customers and poke fun at one another in a way that strengthens the collective consciousness of the group. Sure, many working-class people in the service industry hold peculiar stereotypes about people and unintellectual ideas about the world, but so do many academics—and perhaps especially academics. In fact, one of my greatest shocks about academia was the utterly anti-intellectual environment of much of the university. The academic and intellectual are two different animals—one driven by careerism and insecurity and the other by a joy of discovery and an intellectual curiosity. I have found that both types despise the other, perhaps understandably so, because each threatens to destroy the other's university career. After working in the service industry, I suspect that the service-industry workers of New Orleans hold more intellectual curiosity and joy of discovery than the mainstream academic, despite the toil and monotony of their service-industry trade.

Working in the service industry of a high-volume restaurant or bar is traumatizing. The mind races for hours to keep up with the demands of hundreds of relentlessly thirsty tourists. At the end of the shift, the mind still races, unable to slow down, releasing all its pent-up energy. You are simultaneously exhausted from intense toil yet too stimulated to move on to a new productive task. You need to medicate, and medicate immediately—pot, whiskey, beer, anything, everything. Almost every night after work, sheer exhausted wrecks, we would drink until the liquor took effect. Then we'd catch a few hours of sleep before the next day of toil. These jobs paid us just enough to keep living to continue working. As one service-industry worker so eloquently explained it to me:

> Four-thirty a.m. That's the time I like best. When the last drunks have finally heeded my calls and heaved themselves from their barstools, straggling out the door, and the jukebox and the dozen TVs are turned off. The silence is so loud it thumps in time to my heart pounding in my aching feet. And though I want to go home terribly, I move slowly simply because I can. Pour myself a beer, clean this wretched place up. Wipe the bar down, pull out the barstools, turn off the lights, stock the coolers, clean the popcorn machine. All with a beer of my own in hand and nothing—absolutely nothing—going through my mind except a repeating spool of thoughts that have forcibly consumed me all night: two Bud Lights, get his change, they want shots, what goes in a Jonestown again? Clock out, smooth the crumpled bills that fill the tip jar, calculate how much my toil has earned me today. Turn my car on but not my radio because the silence is what I've craved all night without even knowing it. Collapse into bed with the knowledge that tomorrow will be no different, but yearning for the sweet release that is sleep. And then the dreams come.

This is the reality for thousands of people who work in the tourist and service economy.

OTHER DOWN-AND-OUT JOBS IN NEW ORLEANS

Besides bartending on Bourbon Street, I worked other jobs to supplement my meager income.

Maid Service

My friend Shane introduced me to the lovely Kesha, a friend of mine from Brooklyn who submanages Airbnb listings in some of the neighborhoods surrounding the French Quarter, including a double-shotgun house in Faubourg Marigny. Business was slow, and she wanted to leave town, but there was no one else to clean the apartments between each new arriving tenant. We worked out a deal where I would live on one side of the shotgun house for $500 a month, a cheap price for a place so close to the French Quarter, provided I cleaned the adjacent apartment between tenants. I scrubbed toilets, swept and mopped floors, wiped counters and furniture, washed and dried sheets and towels, made beds, and oversaw the overall maintenance of the apartment, including purchasing and installing an air conditioner.

Miming

After becoming well acquainted with a street mime known as Gold Man—a nickname derived from the color of paint with which he coats himself, head to toe—I learned how to mime as a New Orleans Saints horror clown on both Decatur Street and Poydras Street as well as near Champions Square by the Superdome. Although I received tips, my wages were meager, perhaps three dollars per hour. Through Gold Man, I became familiar with other street mimes throughout the French Quarter, including Uncle Louis and Kenny the Silver Cowboy, as well as other aspiring mimes that Gold Man sometimes informally trains.

Stomping on Glass

Through my contacts with the sideshow "freak" performers Eric Odditorium, a sword swallower, and Stumps the Clown, I found a way to perform as a glass-stomping clown at the bar Mags 940 in Faubourg Marigny.

Writing Poetry

Shane, the man who seems to connect all the new souls in the city of New Orleans, hooked me up with some of the sidewalk poets of Frenchmen Street, who sell their verses to customers strolling along. I became well acquainted with the poet and Bourbon Street "whip girl" Shannon, who walks around with a whip, spanking paddle, and pasties on her exposed breasts. There I wrote poetry for tourists on commission. I also met with other street poets in and around the French Quarter, including the locally known Cubs the Poet.

Selling Pot

I'm a horrible drug dealer. I even tried weighing pot using a smartphone app. Drug money bought me a few beers and a nice evening dinner, but suffice it to say, I'm now retired from employment in that segment of the informal economy.

URBAN LIVING ON THE CHEAP: STRATEGIES FOR SURVIVAL IN THE LATE-MODERN METROPOLIS

No place to stay? No problem.[9] One can camp for the night in one of the scores of vacant houses left abandoned after Hurricane Katrina. The more savvy can check census data online for free to figure out which areas of the city have the most vacant houses. Forget the car—the best way to learn the city is on two nonmotorized wheels. Sleeping in vacant housing's not so bad if you can handle living without the comforts of heat, air conditioning, and running water—and with creepy-crawly cockroaches and the possibility of strangers interrupting your sleep, among other things. You can learn to Dumpster dive for food, clothing, and cardboard boxes, which make sleeping on hardwood floors a bit more tolerable. Urban campers and squatters use plastic bags as toilets, and "bum baths," where homeless people wash themselves in restaurant and bar bathrooms, are available in many of the dingy all-night bars of the city. My friend Lightfoot walked into the bathroom of Checkpoint Charlie to find a hairy, long-bearded homeless man standing sopping wet and stark naked near a sink of running water. They stared at each other, each startled by the other's presence. After a few moments, the man bellowed out in frustration, "Soap, soap, there's no fucking soap in here. Where's the fucking soap?" A strange situation—but quite a reasonable request.

It's not too hard to get a job in New Orleans, especially in the service and tourist industry. Black and Hispanic folks generally get the worst jobs, such as washing dishes and mopping floors; the better-paying positions go to upper-middle-class Anglo-European male cultural capital. It's easy to create a résumé with all sorts of education and work-experience embellishments. Employers in the service industry rarely ask for transcripts. That's because academic credentials and work experience mean nothing. The most important skill is getting the job done. You gain respect in the industry not from fancy papers but from how well you produce under fire and how well you look out for your fellow colleagues. One can make some money in New Orleans with a little "service-industry charisma" and a bit of lagniappe to match.

After hours, service industry workers SIN away (Service-Industry Night) with cheap drinks for those who toil in the bar and restaurants and other tourist jobs. Bartenders and waiters also hook one another up with free and cheap beer at one another's establishments. They consider it a well-earned supplement to their low wages. Owners consider it theft, but of an expected variety. The more drinking-inclined workers develop strategies such as "bar in the car," where an ice chest full of alcohol—purchased at a gas station or grocery store at much lower prices than what is charged at bars—allows bar-hopping revelers the ability to refill their to-go cups between bars. So long you buy the occasional drink at the bar—and tip the bartender—no one seems to notice or care.

COMPARING ORWELL'S 1929 PARIS TO MARINA'S 2016 NEW ORLEANS

Written with Dr. Michael Haupert, professor of economics at the University of Wisconsin–La Crosse. The shift in tone and references to Marina in the third person are results of this collaboration.

The research conducted considers the economic conditions of George Orwell's 1929 Paris and compares them to Marina's life in New Orleans.[10] To make the comparison as close as possible, the researcher Marina must live on the urban fringes of New Orleans, as Orwell did in downtrodden Paris, for ten weeks.

Orwell reported a monthly income of 500 francs in 1929. In 1960, the new French franc replaced these old French francs. In turn, the euro replaced the new French franc beginning in 1999. The French franc was completely phased out of use in 2002 in favor of the euro. Consequently, converting the Orwellian franc to modern-day references requires a series of mathematical applications. First, the 1929 franc was converted into 1929 dollars using the 1929 exchange rate of 25.54 old francs per dollar. Then the 1929 dollar was converted to current dollars by adjusting it for inflation using the standard CPI adjustment. The CPI is the consumer price index, which is a measure of inflation most commonly used to adjust historical dollar values to present-day equivalents.

Orwell's reported monthly income converts to $138.97 today, which is $1,667.58 per annum. This is well below the U.S. poverty level of $11,770 for a single person. In fact, it is a mere 14 percent of the U.S. poverty level and just 6.2 percent of the average wage of a U.S. worker.

There are, however, some problems with making these comparisons. The most important one is converting a 1929 wage into present-day dollars. Using the CPI is only a rough approximation of the purchasing power of a dollar over time, and the longer the time period under question, the rougher the estimate.

The reason that making price comparisons of this nature over a long period of time becomes difficult stems from the way in which the consumer price index (CPI) is calculated. The CPI measures the prices of a sample of goods at different points in time and then compares those price changes to calculate inflation. If a basket of groceries cost $100 last year and the same basket cost $110 this year, then inflation was 10 percent. The problem in comparing these figures over long periods of time is that what specific groceries actually go in the basket do not remain constant. For example, what goods that Marina purchased in New Orleans were available to George Orwell in Paris in 1929? At one point Marina spent money to repair his computer. Computers did not exist in 1929. And some goods available to both Orwell and Marina have changed dramatically. For example, an automobile in 1929 is certainly not the same as today's automobile. Finally, CPI calculations also assume that the same quantity of each good in the basket is purchased each year. These comparison problems become even more obvious when you mix the passage of time with different cultures. The French culture in 1929 differs greatly from the culture of twenty-first-century America.

A better picture of the relative standard of living can be drawn by comparing each man's wage with that of the average person of his time period. This gives us a glimpse of the relative earning power that each had. We can then compare these relative standards. The 500 francs Orwell earned per month in 1929 was 29 percent of the average household income in France. Comparing his earnings to the average adult income at the time, he was earning 50 percent. Although there was no poverty-level calculation at the time, we can put these relative levels of income into perspective by looking at the modern-day American equivalent. Today, the poverty level in the United States for a single person is $11,770, which is 23 percent of the average U.S. household income and 44 percent of the average wage of an individual worker. These ratios are quite similar to the ratios of Orwell's earnings to the average French worker. In other words, Orwell was earning in 1929 about what a single person at the poverty level would be earning today.

Marina's earnings of $3,312 for the total eleven-week period (one more week than Orwell) were 21.6 percent of the average earnings for a 35–44-year-old American in 2015, which puts him at a relative earning level just below that of the poverty line for a single person. Orwell was earning at a slightly higher relative level (29 percent) than was Marina (21.6 percent). However, Orwell's earnings also included a daily ration of wine and meals on the job. If we had enough information on the value of his food and drink, we could convert that to a cash equivalent and add it to his income. Because his income is underreported in this way, it actually puts Orwell in an even more favorable situation relative to Marina. While Marina did receive some free meals, he was not given the equivalent of two liters of wine per day (as Orwell did). Thus, even if we cannot precisely measure it, Marina's relative standard of living can quite reasonably be estimated to be lower than Orwell's.

Orwell lived down and out in Paris for about ten weeks. He begins his book in Paris with 450 francs ($126.87 inflation-adjusted present-day USD) and earns 36 francs ($10.15 USD) a week giving English lessons prior to becoming down and out.[11] He paid in advance 200 francs ($56.39 USD) for a month's rent. Orwell planned to live off of the remaining 250 francs for a month ($70.48 USD), along with additional money from giving English lessons. The plan was to find more work prior to running out of money. He was robbed and left with 47 francs ($13.25 USD) and then forced to live on six francs ($1.69 USD) a day for a month. Once the 47 francs were gone, after three weeks, he was left to live on the 36 weekly francs from English lessons, or 5.14 francs ($1.45 USD) a day. As additional income, he pawned clothes and a suitcase for 70 francs ($19.74 USD) and received 200 francs ($56.39

USD) from a newspaper article. With this, he paid another month's rent. Later, he received a surprising 50 francs ($14.10 USD) for two overcoats in a cardboard suitcase. Finally, with the help of his friend Boris, he found low-wage jobs in the restaurant industry, where he worked for the rest of his days in Paris before leaving for London. He earned 500 francs a month ($138.97 USD), plus free food and two liters of wine a day, as a *plongeur* working eleven- to fourteen-hour days, six days a week.

We compare the costs of living in New Orleans with 1920s Paris using clues found in Orwell's *Down and Out in Paris and London*. Table 3.1 uses hints and information from Orwell's experiences to provide a general idea on the cost of living he experienced in 1929 Paris. Converting 1929 francs to its current value in U.S. dollars and adjusting for inflation, Orwell paid $45.80 for rent in Paris's Latin Quarter. A pound of potatoes set him back 2 cents; a pound of bread cost 23 cents. He could get a half-pound of bread with garlic to rub on it for 16 cents. Orwell could have a cup of black coffee for a mere 11 cents and an alcoholic drink at a bar for 13 cents.

In down-and-out New Orleans life, coffee (with an expected tip) costs $4; the average price of a beer costs about $4 or $5 a pint at most bars in the areas in and around the French Quarter. A loaf of French bread sets you back $3.29 (taxes included); a pound of potatoes costs $1.07 after taxes. The average rent in the city's historic center, including the French Quarter, Warehouse District, St. Charles Avenue corridor, Mid-City, and Downtown, averages $1,291 a month—from $926 for a studio apartment to $1,745 for a three-bedroom and two-bathroom unit.[12] These prices are well above the $500 a month Marina found, through his contacts, in areas close to the French Quarter. Of course, Marina's "cheap rent" sometimes involved sleeping on air mattresses without privacy in dining-room areas or performing maid services in Airbnb listings.

Over an eleven-week period from May 19 to August 4, Marina earned about $301 a week, mainly bartending on Bourbon Street, an amount that includes weekly paychecks as well as cash and credit-card tips. Marina's salary was 21.6 percent of the average salary of an American between the ages of thirty-five and forty-four years of age. Table 3.2 shows each category relative to total expenditures and total income.

Most of the money Marina spent in New Orleans went toward food, alcohol, and coffee. Since Orwell paid rent from income generated prior to

TABLE 3.1 Cost of Living in Orwell's 1929 Paris

ITEM	1929 PRICE IN FRANCS	1929 PRICE IN DOLLARS	CURRENT PRICE, ADJUSTED FOR INFLATION
Bread and chocolate for two	f1.5	$0.059	$0.29
Amount that lasted Orwell for two weeks	f60	$2.35	$13.13
Bread, potatoes, milk, and cheese	f6	$0.235	$1.40
One pound potatoes	f0.1	$0.004	$0.02
One month rent in Latin Quarter	f200	$7.832	$45.80
One drink at a bar	f0.6	$0.023	$0.13
Half-liter milk	f0.8	$0.023	$0.18
One pound bread	f1	$0.039	$0.23
One kg potatoes	f1	$0.039	$0.23
Cup black coffee	f0.5	$0.039	$0.11
One pack Gauloises Blue	f2.5	$0.098	$0.59
Bowl of soup and three rolls	f2	$0.078	$0.51
English lessons per hour	f20	+$0.783	+$5.73
Bus fare	f0.75	$0.029	$0.15
Postage stamp	f0.5	$0.02	$0.15
Newspaper	f0.25	$0.01	$0.07
Cigar	f0.5	$0.02	$0.14
Half-pound bread with garlic to rub on it	f0.6	$0.023	$0.16
66–84 hours/week as a *plongeur* per month (plus meals and two liters of wine per day)	f500	+$19.58	+$138.97

Source: Derived from A. B. Atkinson and T. Piketty, eds., *Top Incomes Over the Twentieth Century: A Contrast Between Continental European and English-Speaking Countries* (New York: Oxford University Press, 2007); and the World Wealth and Income Database, http://www.wid.world/#Database. Inflation adjustments made using CPI. Exchange rate and CPI data from http://measuringworth.com.

TABLE 3.1 Down-and-Out New Orleans Expenditures and Income

CATEGORY	FOOD[a]	ALCOHOL[b]	COFFEE[c]	TRANSPORTAION[d]	ALL INCOME[e]	LAUNDRY[f]	ALL TAX[g]
Total expended	$766.44	$1,343.42	$240.44	$72.14	$3,316.65	$20.24	$62.34
Percent of total expended	27.3 percent	47.8 percent	8.6 percent	2.6 percent	117.9 percent	0.7 percent	2.2 percent
Percent of total income	21.6 percent	37.9 percent	6.8 percent	2.0 percent	93.5 percent	0.6 percent	1.8 percent
Average per week	$69.68	$122.13	$21.86	$6.56	$301.51	$1.84	$5.67
Average U.S. person between 35 and 44 years of age							
Amount	$155.06	$9.71	$8.48	$13.13	$1,394.13	$3.19	$223.06
Percent total expended	12.9 percent	0.8 percent	0.7 percent	1.1 percent		0.3 percent	18.6 percent
Percent after tax income	11.1 percent	0.7 percent	0.6 percent	0.9 percent		0.9 percent	16.0 percent
Expenditures as a percentage of the average American between the ages of 35 and 44							
	44.9 percent	1257.6 percent	257.7 percent	49.9 percent	21.6 percent	57.6 percent	2.5 percent

[a] Food purchased in retail establishments, grocery stores, and restaurants.
[b] Alcohol purchased in retail establishments and bars.
[c] Nonalcoholic beverages.
[d] Public and other transportation.
[e] Total income working as a bartender on Bourbon Street, as well as other odd jobs.
[f] Laundry and cleaning supplies.
[g] All personal taxes (contains some imputed values)

Source: Derived from A. B. Atkinson and T. Piketty, eds., *Top Incomes Over the Twentieth Century: A Contrast Between Continental European and English-Speaking Countries* (New York: Oxford University Press, 2007); and the World Wealth and Income Database, http://www.wid.world/#Database. Inflation adjustments made using CPI. Exchange rate and CPI data from http://measuringworth.com.

becoming down and out, rent in New Orleans was not included in the costs of living. Orwell said, "I was hardly ever without a roof" (23), but Marina often slept under a functional roof (with the exception of squatting in a vacant house and sleeping at a homeless shelter for purposes of the research); he also participated in a few nights of heavy drinking that lasted well beyond dawn.

Of Marina's total down-and-out New Orleans income, 21.6 percent went to food, a whopping 37.9 percent went to alcohol, and another 6.8 percent went to coffee. All other expenses were relatively minimal. Comparatively, Orwell spent 2.6 percent of his income on food and .3 percent on alcohol. While 40 percent of his income went toward rent, that income was earned from labor performed before becoming down and out. Looking at the expenditures as a percentage of the average American between the ages of thirty-five and forty-four is also telling. Marina's total expenditures were 44.9 percent of the average American in his age group. Marina's expenditures on alcohol were an impressive (or frightening) 1,257.6 percent of the average American. Marina's consumption of coffee was more than 250 percent of the average American; his expenditures on transportation were half what the average American in his age group spends. And just as someone stole money from Orwell, someone stole Marina's bike, limiting his transportation to catching rides with contacts or walking.

THE ART OF LIVING IN THE POSTMODERN CITY

In the Rabanesque view of the metropolis, living in the city is an art that requires urban dwellers to channel their own creativity in order to carve out a life. The classic sociologist Georg Simmel reminded us that the great tragedy of modern life is the absorption of our own individual subjectivities into an all-encompassing cannibalizing objective culture. Combining Raban's and Simmel's insights, the city might threaten to swallow and consume us, but it also affords us the space to be remade. The city kills us and allows us to be reborn. It destroys our subjectivity but lets us to build a new one. It pulls our identity into the homogeneous mass but gives us a chance to rescue ourselves from that mass. And in a world riven with inequalities, where the globally transgressive relationship between capital and labor involves oppression and extreme disparity, where chattel slavery is the bedrock of the infrastructure of our society, the task of freeing ourselves and rescuing our subjectivity becomes increasingly difficult. How do down-and-out urban dwellers find dignity and respect in a world in such short supply of those things? How do the down and out find new and creative ways, new transgressive subworlds, that can respond to the collectively experienced structural problems posed by the late-modern metropolis?

CHAPTER 4

BUSKERS, HUSTLERS, AND STREET PERFORMERS

Then the question arises, Why are beggars despised?—for they are despised, universally. I believe it is for the simple reason that they fail to earn a decent living. In practice nobody cares whether work is useful or useless, productive or parasitic; the sole thing demanded is that it shall be profitable. In all the modem talk about energy, efficiency, social service and the rest of it, what meaning is there except "Get money, get it legally, and get a lot of it"? Money has become the grand test of virtue. By this test beggars fail, and for this they are despised. If one could earn even ten pounds a week at begging, it would become a respectable profession immediately. A beggar, looked at realistically, is simply a businessman, getting his living, like other businessmen, in the way that comes to hand. He has not, more than most modern people, sold his honour; he has merely made the mistake of choosing a trade at which it is impossible to grow rich.

—George Orwell, *Down and Out in Paris and London*

PEOPLE BUSK AND HUSTLE all over the place, in nearly every nook and cranny of the Vieux Carré and Faubourg Marigny.[1] Although parts of the area suffer from increasing Disneyfication, these neighborhoods still provide a sustainable habitat for buskers and street entertainers of all sorts. Some people busk to survive; others do it to supplement meager wages earned working in the service industry. Still others busk to avoid the boredom of mainstream, nine-to-five jobs. Many hipsters exploit welfare programs like the Supplemental Nutrition Assistance Program (SNAP; formerly known as food stamps) and busk to avoid working jobs in the traditional economy. Some people busk, as in the case of many brass bands in New Orleans, as a starting point to launch their careers in the local—and if they're lucky—national music industry. Buskers develop a particular skill and take it to the streets. Street performers busk poetry, music of a wide assortment of genres, portraiture and landscape painting, tarot-card and palm reading, miming, juggling, freeze posing, politics, tap dancing, shticks such as "dog playing dead lying in a coffin," comedy, and much more.

Walking west toward Canal Street on French Market Place near the corner of Ursulines Street, where the sun shines directly upon the heads of hundreds of flâneurs, behind the French Market an old man and a middle-aged black woman with a genteel smile play simple classic folk tunes under an umbrella. Next to them is a sign reading: "If you like the sound, stick around, if ya gotta split, leave a bit." She yells to one man taking a picture, "How about a tip? You got more money than me, sir." Nearby, wearing a star-spangled cowboy hat, the political activist and conspiracy theorist Michael DiBari sits next to his Roving Info Wagon and busks radical politics, sometimes arguing with "uninformed" tourists. A board sitting on his cart reads: "If you can change one thing in government, what would it be?" DiBari uses his graphic-design skills to create political postcards and posters with pictures: One depicts President Obama as a vampire slitting the throat of the Statue of Liberty while stealing her torch. Another depicts a corporate-owned news network image that reads, "Con news today: More shit we made up to socially-engineer you fuckers." He hopes to add to his mobile political cart a glass case and enhanced video equipment, so he can interview city officials and others about local and national politics.

Further down the road, a young black kid drums on repurposed white buckets that once held drywall compound; another younger kid watches, bored, from his bike. A few yards down, eight young black kids, middle- or high-school age, blow brass to a large audience next to the Flea Market. They play sousaphone, trumpet, tuba, sax, and drums—the usual musical suspects in this thing called funk brass jazz. Many youths from the surrounding historically black and poor neighborhoods harbor hopes of one day playing at the famous Uptown music venue Tipitina's.

Heading north, in front of a shop calling itself Wicked Orleans, a man with long brown hair tucked under a cowboy hat sits in front of his Harley, money spilling out of his open guitar case. He's singing old-school country music and tapping on a bass-drum pedal. His guitar amp hangs on his motorcycle. Another amp, connected to his microphone, sits on the ground near his drum.

On Ursulines and Decatur Streets just past the saloon Molly's on the Market, the Slick Skillet Serenaders play on the corner as passersby drop dollar bills into their tip bucket. Nearby, two gutter punks sleep next to a sign reading, "Got a dollar? Beer now and pop tarts in the morning." Another sign reads, "Gotta dime for some slime?" Their dog seems bored.

Next to them, a middle-aged black woman sells water for a dollar, and another gutter punk sits in his filth, wearing a jester hat and petting his two dogs. On Rue Decatur in front of the Artists Co-op, seven youths take a break from busking, smoking cigarettes in the shade of an old tree. One of these young musicians performs at various venues in Faubourg Marigny, ones that often charge a cover and attract large crowds. On the corner of Decatur and St. Ann, two dirty gutter punks in rags sit with a cat and dog. One of the punks rubs his tattooed face all over the dog. Tourists, waiting in long lines for a beignet at the Café du Monde, stare.

Across the street, a young black kid no more than six years old, with tacks hammered into the soles of his Nike Air Jordans, tap dances for the fascinated tourists. He claims to make three hundred dollars a day and says he buys his own shoes, of which he says he has many pairs. A cursory observation suggests he averages six to seven dollars for each thirty-second tap-dancing performance—well above the average wages of many Americans. His pockets can barely hold the money. A white male tourist with a thick country accent condescendingly orders the boy to dance faster. He drops five bucks into the bucket and walks away.

Just a few feet away, a limber black kid in his early twenties, covered in grey body spray and hair painted gold, makes his body roll like a moving escalator as he mimes to hip-hop electronic music in front of a large attentive crowd of about sixty tourists. His performance lasts about three minutes.

Tourists give him a hearty ovation as they drop money into his can; at least one man drops a ten. The kid does this all day and easily makes three or four hundred dollars for his efforts.

A middle-aged black guy sits in front of the Café du Monde, facing the beignet-munching tourists while playing a jazzy rendition of "Hello, Dolly," Satchmo-style. His trumpet case holds four dollars in tips received from passing tourists.

On the other side of the French-colonial Place d'Armes, now Jackson Square, a man paints caricatures of musicians leaning on the street signs of New Orleans under bright orange sunlight. Dozens of artists display their work on the wrought-iron fence encircling the square. Tarot-card and palm readers line the pedestrian walkways, sitting next to tables and ready to serve curious customers. A striking woman at the "Henna by Jenna" table tattoos tourists with Indian-inspired designs in the shadow of the centuries-old St. Louis Cathedral, the jewel of the square. Eric Odditorium, a street clown with a tattooed face, swallows swords for freaked-out tourists.

Further back, in front of the cathedral, yet another funky brass band composed of about eight black youths attracts a huge crowd of tourists. The crowd remains standing for more than an hour as these eight young men play their hearts out. This is how the brass bands of New Orleans get their start. One kid walks around with a brown cardboard tip box. The tourists tip these musicians well.

Royal Street becomes a pedestrian mall at Orleans Avenue, closed to vehicular traffic. In front of the Rouses supermarket—a popular busking spot—four brass-playing buskers crank out old New Orleans tunes. It's here that the renowned New Orleans musician Troy "Trombone Shorty" Andrews got his start.[2] One guy puffs into a tuba while keeping the beat with cymbals. Before them are six white buckets, each emblazoned with a money sign. A half-naked man walks past with a crawfish on a leash.

Further west toward Canal, on Royal and Toulouse, a black couple plays Caribbean-influenced music near a sign stating: "Quit our corporate jobs to do this full time." The man sings and plays guitar while the woman concentrates on vocals. Further up the road, Rastafarians with a television camera pack up their equipment on St. Louis and Royal streets after busking. An older man with thinning gray hair pulled back into a ponytail plays acoustic guitar in front of Café Beignet. Nearby, a man makes flowers shaped from palm fronds. A dog wearing a black-and-white tuxedo and holding a tennis ball under his left arm "plays dead" in a coffin next to a sign that reads: "Need money for a proper burial. If it's worth a picture, it's worth a dollar," attracting the attention of a cooing audience.

Royal Street between Iberville and Canal serves as a skid row, with tattooed homeless men drunkenly stumbling about. A black guy plays the harmonica, hoping to attract the attention of a group of young professional-looking female tourists. Eight gutter punks sit and beg by the side of the McDonalds. A thin, young female gutter punk—walking with two older men—suddenly stops, drops her pants, squats, and takes a shit in plain view.

Heading east on Bourbon Street, the strip joint called the Hustler Club welcomes you to that notorious street of unapologetic sin and debauchery. Thin, nearly naked black and Hispanic women in exotic lingerie stand at the club's thresholds, trying to lure randy and leering tourists. A bouncer barks, "Don't be shy, get some titty in your eye," to entice the wide-eyed tourists. Two hustlers approach a white couple and point to the man's shoes: "Twenty dollars I can tell you where you got dem shoes at," a timeworn New Orleans street con. (The answer, a play on the local vernacular: "You got them right here, on your feet on Bourbon Street, in New Orleans, Louisiana.") The man replies, "Man, I'm too old for that shit." One of the hustlers tries the same thing on me, to which I reply, "I'm from Gentilly." They respectfully nod and set off to find another sucker. A man approaches me to tell his story of homeless shelters and how rife they are with theft and homosexuality. He asks to use my phone.

On the street adjacent to the Royal Sonesta Hotel, a band plays while drunken tourists dance. This particular group of musicians are the same ones that earlier took a break in front of the Artist's Co-op. A black man with a worn, wrinkled face, wearing a pink brassiere, pretends to climb a well-balanced ladder. He yells at people to give him a dollar for the pictures of him they've just taken. Steve, a former certified welder and current squatter in a vacant house, stands next to his dog, Eugene, who is lying on his back and pretending to be passed out, two New Orleans hand grenades—the name of a popular French Quarter cocktail—tucked under his front legs. "He's my best friend," Steve says. "We both rescued each other." An amused tourist asks, "Does he move if you tickle his balls?" A man dressed as the animated character SpongeBob SquarePants waves to tourists on Bourbon and Toulouse.

The ersatz SpongeBob says he wants to kick Minion's ass, another man in a cartoon costume, who sometimes stands on that very same corner.

Three black kids younger than ten years old tap dance on Bourbon Street in front of Pat O'Brien's Piano Bar, "Home of the Famous Hurricane." One screams out "tip, tip" to the people taking their pictures. Even at such a tender age, they have developed the art of the hustle.

Consider some of the terms used to describe the poor. Lazy and dull? Lacking in industry or creativity? Pathological and undisciplined? Welfare dependent? Bullshit. I think about how some of my university students demand strict guidelines and require detailed instructions on assignments and papers. How many milksops from middle- and upper-class families could find creative ways to make something out of such dire straits? The poor do it every day in these New Orleans streets. They hustle and they tap dance. They find a way. In a world where state-sponsored capitalism stifles competition for all but the elites, democratic equal-opportunity capitalism flourishes on the streets of New Orleans.[3]

On Bourbon and St. Peter streets, a guy "freeze poses" in a New Orleans Saints jersey, football in hand, hoping for tips. The iconic New Orleans mime known as Tim the Gold Man threatens to kick the football guy's ass, accusing him of stealing his pose. A block up the road, more young kids of about seven years of age tap dance for money. A child who appears to be no more than four runs up to tourists, demanding money for the entertainment provided. A lonely tarot-card reader with sad eyes sits under a big yellow umbrella, waiting for customers on Bourbon Street and Orleans Avenue.

On Frenchmen and Chartres, a traditional New Orleans second-line parade ends, and a brass band jams in front of about a hundred fascinated tourists, who are dancing wild on the street—New Orleans style.

Cocaine-fueled polar bears wearing nipple clamps and smoking Cuban cigars zipping down Bourbon Street on roller blades while chasing whiskey-addled panda bears in stiletto heels would barely catch the attention of a native New Orleanian. To the tourists, second lining and street dancing excites their senses and allows them to relax "their sphincters,"[4] maybe for the very first time in their lives. One block away, just past the music club d.b.a., writers in fancy fedoras and John Lennon–style glasses type poetry about the busker blues while another brass band plays its heart out, hoping for a chance to become famous.

BUSKER SPOTS OF THE VIEUX CARRÉ

Jackson Square, the French Market District, and Royal and Bourbon streets serve as the main busker spots in the Vieux Carré, or what most people know as the New Orleans French Quarter.

Jackson Square

Once called the Place d'Armes, Jackson Square sits along Decatur Street between the Jax Brewery Shopping Mall and the French Market, in front of St. Louis Cathedral and across the street from Café du Monde. For a few dollars, tourists can jump on a mule-drawn carriage in front of the square for a ride in and around the Vieux Carré. Buskers do not need a license to perform directly in front of the cathedral. As a result, dozens of artists, street performers, musicians, and readers share this space throughout the day, making a living from the city's streets. The venerable apartments overlooking the square house a wealthy class of people. The city's noise ordinances keep the sound level on Jackson Square below an eighty-five-decibel limit. Buskers work the corners of Jackson Square but leave the sides directly in front of the Pontalba Apartments to the artists, who hold difficult-to-obtain sales permits. City ordinances allow buskers and musicians to work on the benches on the square but keep them from performing on the steps leading to the cathedral and the grassy park that occupies the center of the square. Buskers must set up at least twenty feet from the black fence surrounding the park area of the square and must not block entrances. All music must stop at 8 p.m.

French Market District

The French Market District includes the areas north and south from the Mississippi River to Chartres Street and west to east from the shops in the Pontalba Buildings near St. Peters Street to the Old U.S. Mint near Barracks Street. More than two hundred years old, the French Market began as a Native American trading post on the banks of the Mississippi River. The French Market District includes the Farmers Market, where vendors sell local produce on Wednesdays and Saturdays, and the Flea Market, where locals and tourists alike can haggle with vendors selling everything from leather belts and wallets to fleur-de-lis earrings. Strolling through the French Market District includes a visit past the great statues of Dutch Alley and the converted row houses of the elaborate, cast-iron filigreed Pontalba Buildings, which are described as the "oldest, in some ways most somberly elegant, apartment houses in America."[5] This is where locals scream "Stella" and "Stanley" during the Tennessee Williams New Orleans Literary Festival and where the literary great Sherwood Anderson once lived. The French Market District also includes the Moon Walk promenade along the Mississippi Riverfront and Washington Artillery Park, offering the best views of the cathedral, which many believe is still haunted by its former priest Père Antoine; and Latrobe Park, which used to function as a mini skid row where gutter punks used to hang out, until French Market authorities "cleaned it up."

The French Market hosts many events and festivals throughout the year, and these attract huge influxes of locals and tourists, from the Mardi Gras Mask Market in February to the Creole Tomato Fest held in June to the Boo Carré Halloween and Harvest Festival in October. Even without the festivals, the dozens of shops and restaurants of the French Market District, including the world-famous Café du Monde, produce a stampede of foot traffic to the area throughout the day. It's a busker's paradise, and getting a license to perform on the street takes almost no effort at all. The French Market Marketing Office offers street performers and buskers free licenses. They issue badges to buskers, allowing them to perform on the streets legally on a ninety-day registration cycle.

Aside from a few rules concerning issues such as using speaker amplification, animals, and potentially hazardous materials such as fire, street performers have a legal right to perform without interference from city authorities, including from French Market District security. It is also important to note that buskers cannot legally solicit people for money or sell anything. Rather, street performers may place donation baskets or buckets, or what some buskers call "buskets," near their performances. Buskers must compete for the most desirable spots in the French Market District; only four performers at a time are allowed on a spot. Because of noise complaints, brass-band musicians are allowed to perform only between Dumaine Plaza and Washington Artillery Park in the French Market. Aside from these rules, the French Market District serves as a prime urban busking spot given its large and steady flow of tourists throughout the day, every day.

Royal and Bourbon

Although city ordinances define and regulate the street performers of New Orleans, most buskers find it easy to avoid problems with the police. Bourbon Street draws rowdy tourists from all over the globe. Although the street remains relatively quiet in the early afternoon, Bourbon begins to attract huge numbers of tourists as the evening approaches. This is perhaps one of the best busking spots in the world, though for some buskers it is sometimes dangerous and unnerving. According to a city ordinance, street performers must cease their activities between 8 p.m. and 6 a.m. on Bourbon. The more tranquil and sophisticated Royal Street, which runs parallel to Bourbon one block closer to the river, hosts antique shops and art studios and offers another excellent busking and performing area. The Rouses supermarket on Royal and St. Ann serves as one of the hottest and most cherished busker spots in New Orleans. At certain times of the day, city officials close the street to vehicular traffic to increase pedestrian space in specific areas on both Bourbon and Royal, making these two streets prime busking real estate.

NEW ORLEANS AS A SUSTAINING HABITAT FOR BUSKERS

The Vieux Carré offers a sustaining habitat for buskers to make a living performing on the streets of New Orleans. The main qualities of such a habitat include (1) heavy foot traffic in tourist hot spots and pleasure-seeker zones; (2) busy bars, restaurants, shops, and music venues; (3) short-term tourists with money to spend; (4) easy-to-obtain busker licenses; (5) laissez-faire attitudes from city authorities about street performers; (6) urban sidewalk spaces that do not block walkways and banquettes; and (7) sympathetic business owners who do not complain to rule enforcers. For the most part, with some exceptions, which will be described below and in the next chapter, street performers enjoy a busker's paradise in the Big Easy.

The city's noise ordinances typically prohibit public performances after 8 p.m., but this does not seem to apply to—or at least is not enforced on—Frenchmen Street (see the next chapter). Sometimes the rules for street performances are stringently enforced, especially during large festivals. At other times they are largely ignored. Sometimes, business owners complain about the buskers, calling police and claiming that buskers interfere with their business. Buskers must find urban spaces that do not conflict with the interests of business owners. As a result, they search for locations in the ambiguous urban spaces between businesses to perform. Other businesses, like the Rouses supermarket on Decatur and St. Peter, rarely, if ever, cause problems for street entertainers. Such "friendly" business owners still retain the right to ask buskers to leave or risk a call to the police.

The Vieux Carré offers buskers all the resources necessary for daily survival: ample access to bathrooms in the many bars, cheap drinks in many of the small grocery stores, and easy access to illicit drugs, with local dealers serving both local and international clientele. The buskers know one another well, and while some disputes emerge with regard to turf and the copying of styles, most get along. Some buskers, like the famous mime Tim the Gold Man, help budding street performers, teaching them their craft, even if it adds to the competition. Further inspection of the city's busker community reveals a network of street performers informally offering one another support. Some of the performers live together in places like the "Clown House," where the sword swallower Eric Odditorium and Stumps the Clown once lived and where Tim the Gold Man currently resides. Buskers will often house people in New Orleans to help them get established or at least back on their feet. Others share the wealth, spotting a few dollars here and there when some members become down—and sometimes very close to out.

Buskers also often provide networks of emotional support as well as opportunities to achieve local fame in the city. What is more, hundreds of tourists who drop money in buskets take pictures of these men and women performing on the streets. Their pictures must be in homes, phones, and laptop computers all over the world, not to mention all over such social-media platforms as Flickr, Facebook, and Instagram.

The buskers of New Orleans exist on the fringes of a multi-billion-dollar tourist economy, often making enough money to survive. This unconventional economic street life serves as one example of the type of creativity that thrives at the edges of the postindustrial capitalist economy. The postindustrial society includes, among other things, widespread economic restructuring in a highly fluid and rapidly changing economy, with increasing inequality between the wealthy and working classes.[6] This includes a shift from manufacturing to service-sector jobs and an increase in marginal jobs that offer low wages and even lower social status.[7] Such growing inequality has created an underclass of socially excluded people who carve out lives on the fringes of our societal institutions.[8] These folks are part of the new underclass, but they live as outsiders and refuse to accept the conditions, or at least the jobs, of the working poor. Some of these street performers struggle to secure regular employment that offers livable wages; others could, if they wished, put on a suit and get an office job any day of the week. Others deliberately exist on the urban economic fringes in a protest based on philosophical differences with mainstream society.

Although some make only a few dollars, more innovative and talented buskers make hundreds of dollars a day, well more than that earned by many hard-working New Orleans service-industry employees.

Just as New Orleans provides a sustaining habitat for buskers, street performers offer a profit-sustaining habitat for the tourist industry. They entertain and fascinate tourists and enhance the scenes and soundscapes of the city. Buskers and street performers are part of the heart and culture of New Orleans, and they are part of what attracts tourists from all over the world, bringing in big bucks for the tourism industry.[9]

Some street performers play for money as well as for the love of the music and culture of New Orleans. The musician known as Tuba Fats, a former street performer and one of the founding members of the famous Dirty Dozen Brass Band, says, "I don't need to be a millionaire. If I want to play on the street, that's my business. We're not beggars, we're not homeless, I play in Jackson Square and I do it because peoples love music and I love to see peoples enjoy music. People come to New Orleans to hear the music and they don't get it up and down Bourbon Street. It's not there anymore."[10]

"Officer Joe" of the New Orleans Police Department—not his real name, as he wasn't authorized by his superiors to speak on the record—reveals the sentiments that some police officers have, as well as other locals from the community, about people who make a living on the streets of New Orleans. Some people, like Officer Joe, wonder why street performers decide to avoid conventional jobs in favor of working the tough city streets. "They seem lazy to me. They don't want to go to work. They don't want to get a job like everybody else, so they just make an easy living off the streets."

Some innovators find new and creative ways to achieve desired cultural goals when institutional means seem insufficient.[11] Busking and other jobs in the informal street economy is one such creative solution to the poverty wages the New Orleans service and tourist industry offers. Although some believe that street performers are lazy, poor, and lacking dignity, it actually is drive and ambition that allow street performers to make their livings as self-employed entrepreneurs. Yet people still perceive those who work in the transgressive informal economy to be lazy because they refuse to get regular jobs, poor because making money from the streets resembles begging, and lacking in dignity because they exist outside the status-bestowing formal economy.

THE SOCIAL AND ECONOMIC LIVES OF
NEW ORLEANS BUSKERS

On the corner of Decatur and St. Ann Streets, an hour of observation reveals that the young black kid tap dancing with nothing but the top of a coke can nailed to the bottom of his shoes, a milk crate acting as a busket, makes an average of six dollars for every twenty seconds of performance. He tap dances about once every two or three minutes, usually when a huge pack of tourists are about to pass him. He does this throughout the entire day, earning anywhere between two and three hundred dollars. Of course, some of it goes to his parents or guardians, who take it from him once he gets home, but this clever tap dancer hides much of this money in his shoes and underwear. At such a young age, this young kid's rate of income is higher than that of many college graduates.

On one side of Jackson Square close to St. Ann, Tyrone and his brass band play four sets to a large tourist crowd in under an hour. The seven musicians together earn between $350 and $400 at the end of each of the four sets. That's at least $200 for each musician for one street "gig." This gig is but one of a number they play throughout the day. Tyrone sits on the steps of St. Louis Cathedral with his wife and kids. "I love what I do," he says, a big grin on his face. "I love playing music for the people, making them happy. I support my family doing this, got a car and cellphone and nice apartment in New Orleans

East." It's easy to spot many of these musicians later in the evening performing in front of large audiences at music venues such as d.b.a., Blue Nile, Vaso, and Balcony Music Club (BMC) as well as at the large festivals, which attract tens of thousands of tourists. These brass-band members also work on commission for second lines, funerals, memorials, and other celebrations of life and death. Often with no formal training outside of perhaps participation in their high-school marching bands—sometimes even less—and without access to the institutional resources afforded budding musicians in the middle class, many brass-band members start their musical careers busking on the streets of New Orleans.

Many of these tap-dancing kids and young black musicians often skip school to make a living on the streets. New Orleans, and the rest of the United States, fails to provide a reasonable alternative to many young black kids as to why they should give up their street life and attend school to "better" themselves. They can make hundreds of dollars a day performing on the street, well more than the poverty wages awaiting them once high school ends. This reminds me of a story my friend Bill Zollweg, a retired professor from the University of Wisconsin–La Crosse, recalls:

> A sentencing specialist in Michigan feels it his duty to council an eighth-grade male named Jerome who recently skipped 120 out of 180 days of school. The sentencing specialist talks with Jerome in his fancy office about the middle-class virtues of attending school, respecting authority, getting jobs, and obeying rules and laws. The young Jerome, slumped in the chair on the other side of the huge desk, leans across to the sentencing specialist: "You must think that I am stupid. Just because I don't go to school does not mean I'm stupid." Jerome continues, "Listen, I stand on the corner of Sixty-Eighth and Simpson Avenue to blow a whistle every time I see a cop. When I blow the whistle, all the dealers flock to safety. They return once the cop leaves. These dealers pay me $200 a day. Why would I give this up to go to school so that when I graduate I'll work fifty-two weeks a year for poverty wages?" As the surprised reading specialist listens, the counselee becomes the councilor as Jerome confidently points out, "I make more money than most college graduates. I'll probably retire before you." Jerome, now looking deep into the eyes of the sentencing specialist, asks, "Now, who is the stupid one?"

It's a social problem when a society cannot offer a realistic alternative to Jerome's reasoning. Until society offers a viable solution for Jerome, we should expect people to make a living in the informal and transgressive economy of the streets—whether it's selling drugs, busking, or hustling tourists. People need money to survive in a capitalist society—and one way or another, they will get it, finding creative ways when the legitimate paths fail or seem ineffective. Graffiti written on an abandoned house in Mid-City reads, "Housing and homelessness = society fails." It seems like it's failing again for many American citizens and young men like Jerome.

Directly in front of the St. Louis Cathedral, the tattooed sword-swallower Eric Odditorium performs his daily routine, shoving swords down his throat in front of about a dozen tourists. Most of the people in the crowd drop anywhere between one and five dollars, sometimes ten, into his sword case. This act takes between five and seven minutes. He usually performs it between ten and twenty times a day. Eric works roughly from 11 a.m. to 5 p.m. every day, except on the slower Tuesdays, making what he estimates to be an average of $100 to $150 daily during tourist season, although sometimes much less during the summer off-season. He supplements his income by performing in his Cut Throat Freak Show South, with Stumps the Clown, at venues such as Mags on Elysian Fields in Faubourg Marigny.

The beautiful Jenna, of Henna by Jenna, sits behind a table in the cool shadow of the iconic cathedral, ready to tattoo customers with temporary body art. She sets up her makeshift studio at about three or four in the afternoon and works until the sun begins to set. Jenna usually makes about $20 for each tattoo, and she sometimes has long lines of customers, many of them young girls with their families, who keep her busy throughout her shift. Jenna says that she makes anywhere between $100 and $300 dollars in a four-to-five-hour workday, depending on the season.

Uncle Louis makes enough cash every day to continue his "black Uncle Sam walking a dog" routine. He's been performing this routine for about thirty years. The tourists love him, many stopping to take pictures and throw money into his busket. He teaches a tourist how to pantomime with enthusiasm while making jokes about interest in his street performance. He's a blue-collar street performer, working regular shifts to make a living.

Kenny the Silver Cowboy has been coming to this city for decades from Mississippi to pantomime, standing on a crate on Decatur Street. He makes enough to pay for food and rent. It might not be much, but it does just enough.

Michael the Political Busker—the guy with the red, white, and blue cowboy hat and all those conspiracy theories—sits for hours debating with pedestrians while hoping to give away his radical political art for a "donation." Meanwhile, the folk musicians Roselyn and David claim to have supported their kids busking in Jackson Square and the French Market District. They also used their busking money to buy a house in the now posh Bywater neighborhood. They've achieved the busker version of the American Dream. Shannon the Bourbon Street whip girl will "discipline" tourists for a dollar donation. She says she loves her job, which is obvious as she giggles while chasing after tourists who chicken out of the whipping after giving her money. She claims to make a killing as a whip girl, and boasts of the prestige and status she receives from being the hottest act on Bourbon Street.

Soap Man claims to make his own natural soap and creams, displaying his elegant products on a table in front of a closed local business on Decatur Street. He's been busking various hustles for much of his life, from city to city. Soap Man proudly proclaims that he is an entrepreneur who has learned to make a decent living selling products on urban streets. The silver-painted Reuben mimes to hip-hop electronic music; his three-minute performances make him about three or four hundred dollars a day. Two break-dancing kids on Bourbon Street make a couple hundred dollars a night with their street show. They are self-effacing, using a barrage of black stereotypes to make the mostly white tourists laugh, and these young men tear it up, making good money. One of the kids also appeared on the television show *America's Got Talent*. That's a source of pride and status for young men from places like the Ninth Ward.

On Royal Street, Cubs the Poet sits on a crate with a typewriter on his desk, typing away for hours about love and life. "We are all poets," he says. "We all use words, we all have beauty." He writes, he says, because he must— because he is. Cubs was one of the first poet buskers in New Orleans, and, while many poets struggle to publish, he makes a living from his street verse.

The 1920s-style jazz singer Meschiya Lake started busking as "Nurse Nasty," hanging around with crazy circus folks like Stumps the Clown before reaching local celebrity status and playing in popular music venues like the Spotted Cat. "Music saves people," she says. "It saved me. New Orleans is the town that made it happen."[12]

Not everyone fares so nicely. Sheila stands nearly naked, wearing antlers and with exposed breasts, save for her pasties, with a sad look in her eyes; the tourists of Bourbon Street either harass or ignore her. She lost her conventional job months ago. Frustrated with the lack of well-paying job opportunities, she complains there are just not enough ways to make decent money. On one night she sat alone on Toulouse Street, her knees resting on her chin, a trickle of tears falling down her cheeks.

Busking is not for the lazy and unmotivated. People such as Cubs toil every day to improve. Eric Odditorium trained for years to be able to shove a sword down his throat without hurting himself. Musicians work hard to learn how to play well enough to generate a crowd on the streets. Others find creative ways to make money using nothing but their wit. Gold Man, who occasionally works as a grass cutter and house painter, learned how to pantomime without any training, creating a pose that has pulled money into his busket for almost three decades.

It's not just money that street performers receive busking in New Orleans. Many launch their musical careers on these city streets. Others achieve local celebrity status, like Tim the Gold Man, by inventing creative performances. Under other circumstances, people like Tim often fade into obscurity working odd jobs and drinking heavily until the inevitable end. The conditions were ripe for Tim to meet that fate, but he used his agency to beat the odds. He transitioned from a poor, wandering, homeless traveler to a New Orleans minicelebrity. Who dat? Gold Man dat. It is clear the buskers, as a whole, earn wages above the poverty level, work hard to develop their skills and talents, and enjoy status and prestige from performing. In short, the labels often attached to buskers—poor, lazy, undignified—fail to match reality.

TIM THE GOLD MAN

Gold Man says, "I'll be out here till I die; this is my retirement home." It started for him almost three decades ago when he was found drunk on a highway while traveling to, well, he doesn't even remember. He eventually landed in New Orleans and worked as a Lucky Dog vendor, selling hot dogs to tourists and drunk locals, Ignatius Reilly–style, with a little pot dealing on the side as lagniappe.[13] One day, like many days, he drank heavily, and like many good ideas that come from a night's drinking, *Eureka!*

Gold Man: Well, when I came out here, it was 1982, and I watched the
Silver Man. . . . I was dressed in regular clothes. Regular-ass clothes,
dude. So anyways, I ran outta money, right? For beer. So I put an empty
beer cup in the middle of Bourbon Street, in regular clothes. And
I stood on one foot and I pointed to the fucking cup. Man, son of a
bitch. I had thirty fucking dollars in less than fifteen fucking minutes
and Silver Man got mad and tried to whoop my ass. And I'm in regular
clothes, dude, I didn't have no makeup or nothing on me, man.

Marina: So how did that work out with you and Silver Man?

Gold Man: Well, actually we all kinda try to get along, but some people are
just better than other people. See, nobody can keep their eyes open for
an hour. See? No blinky. My eyes are trained to do that.

Marina: Really?

Gold Man: Yeah. Look me in the eye for an hour, you want to try me?

Marina: [*laughs*] How do you do that, though? I believe you, I believe you.

Gold Man: I don't know, bro, it's like yoga. I do yoga, so I control every
fucking part of my body, dude.

Ladder Man in a Bra (another busker): Every part?

Gold Man: Can you do this? Betchya I can show you something you can't do.

Ladder Man in a Bra: What?

Gold Man: Got a dollar?

Ladder Man in a Bra: I got a dollar, I got two.

Gold Man: Watch this [*Gold Man proceeds to do a near full split*].

Gold Man has been performing on the streets of New Orleans for twenty-six
years. He's now part of the social structure of sidewalk life in the French
Quarter—what Mitch Duneier, borrowing from Jane Jacobs, calls a self-
appointed public character.

A public character is anyone who expends the effort and is sufficiently
interested in staying in frequent contact with a wide circle of people. Public
characters need have no special talents or wisdom to fulfill their function—
although they often do. They just need to be present. Their main qualifica-
tion is that they *are* public, that they talk to lots of different people. It is
through the public character that news of sidewalk interest travels.[14]

Gold Man knows just about all the characters that make up the street life
in the Vieux Carré. He knows all their stories and the stories of hundreds

more who have come and gone through the years. "Everyone in this town knows who I am. I know a shitload of people. I treat 'em like I wanna be treated." Indeed, interviewing Gold Man on the streets of New Orleans involved constant interruptions from people wanting his attention. He tells the stories of the talented and iconic Uncle Louie of Bourbon Street and Kenny the Silver Cowboy, who carried a whistle in his mouth, as well as others who have spent years making a living busking.

As we walk down Decatur Street just past Tujague's, the second-oldest restaurant in New Orleans, Gold Man strikes up numerous conversations with seemingly random people, from service- and tourist-industry workers to tourists to street beggars. Sometimes he gives beggars some of the cash he busked that day or cigarettes and marijuana. What is perhaps most surprising is his propensity to teach other down-and-out urban dwellers how to pantomime.

> *Marina:* So you train other people? Why do you do that? It adds to your
> competition.
> *Gold Man:* I don't know, man. Stupidity. They're always a thorn in your
> ass, dude, because, anyway, I tell them certain things to do, how to
> respect me teaching them: "All right, Imma show you the ropes, and
> I'm going to let you go ahead with yourself, because I'm the famous
> one." Been here the longest, twenty-six years.

Performers must follow the informal rules of the street to avoid stepping on someone's toes and creating trouble. They compete for the best spots to busk and engage in constant negotiations that usually keep the peace. They share the best spaces throughout the day, reaching compromises on such matters as how long to perform in a given spot before ceding the space to the next busker. A willingness to share space and to compromise often avoids conflict. The general rule, however, is that space belongs to people grandfathered in—that is, those who have simply been there the longest. And copying someone else's act is a clear violation of a street norm. One day while I was walking with Gold Man down St. Peter's toward Jackson Square to meet Eric Odditorium, Gold Man spotted a man using his pose to busk on Rue Royal. Gold Man first stared at the shtick-stealing interloper, then challenged him to a fight, flailing his arms and yelling, at which point I urged

Gold Man to continue toward the famous sword swallower. The takeaway message was clear: Mimicking another person's street performance is against the rules, and there are penalties for violating the informal street code.

> *Gold Man:* You know how many people I fucking got into a fight with over a spot, dude?
> *Marina:* Really?
> *Gold Man:* Oh yeah. Been to jail thirty-six fucking times, dude, doing this.
> *Marina:* You have? Why?
> *Gold Man:* All misdemeanor shit. See, New Orleans is a corrupt city, and what they got going on is, you got to get a certain amount of arrest credits, alright? In order to get your Christmas bonus. So right around Christmas time, they lock everybody up. All misdemeanor shit, dude.
> *Marina:* Why?
> *Gold Man:* Obstruction of a sidewalk, masking. They call this [*pointing to his gold-painted face*] masking. I said to the cop, "Don't your old lady wear makeup? That's masking, dude, to cover up that ugliness your old lady got." But I got my ass kicked, dude. Boy, did I get my ass kicked.

Aside from making enough money to survive, serving as eyes on the street, entertaining tourists, and teaching down-and-out people to pantomime, Gold Man also finds a sense of personal accomplishment and dignity from the minor celebrity status busking has brought him. That status is reinforced even when he breaks minor laws and rules. A photograph of Gold Man made it into the pages of the local paper because someone took a photograph of him stealing a newspaper from a front porch. As he tells it, he had only enough money on his person for beer but thought, "Why not both?" So he bought a beer and stole a paper. As Gold Man boasts about his headline-making ways, it becomes apparent that he finds a personal sense of pride and notoriety at being a New Orleans icon, a fixture on the streets of the old Vieux Carré.

> *Gold Man:* Yesterday's paper, dude. I been signing autographs.
> *Marina:* Did it say "Gold Man"?
> *Gold Man:* Oh, they knew who it was. Everyone in this town knows who I am.

Gold Man's local fame and prestige is apparent. Many people stop to talk with him on the streets, and there are constant shouts from pedestrians, bicyclists, other buskers, and drivers: "Hey, Gold Man!" "Yo, Gold Man!" The local police definitely know him, given his thirty-six arrests. Aside from that dubious distinction, Gold Man talks about how he started "hitting it big" and getting into movies, magazines, TV commercials, and commercials for the local lawyer Morris Bart, a sort of real-life, law-abiding version of the lawyer character in the TV drama *Better Call Saul*. Although it is difficult to confirm all of Gold Man's boasts, he does appear in Jim Flynn's clever book *Sidewalk Saints*, which offers portraits and brief character descriptions of New Orleans street performers.[15] A traveling misfit drunkard travels to New Orleans almost three decades ago and becomes a local celebrity. A city like New Orleans offers this possibility to all the transgressive street characters of the world.

ERIC ODDITORIUM: THE BUSKING SWORD SWALLOWER

Eric Odditorium's path involved intensive training and dedication to a craft few people aspire toward. Always feeling himself an outsider, he embraced the role of self-made "freak" that could shock and entertain others. "I was the kid in school who was kind of labeled the class clown. I used to do things like poke needles through my skin, eat bugs, and convince people to pay me money to jump off of buildings." Just like any conventional skill or blue-collar trade, Eric worked hard to perfect his craft. "I'd do broken glass stunts, animal trap stunts, eating bugs. After a couple years of that, I kept progressing, started doing fire stunts, pin stunts, eventually eating fire then blowing fire, which I hardly ever do any more. I still will do 'human pincushion,' which includes self-piercing and deep tissue piercing. You take skewers and pierce through your arm, through your jaw, through your cheeks."

Eventually, Eric began the process of learning to swallow swords, which required years of practice to control his gag reflex and the two sphincters that keep the contents of the stomach where they should be. All of this involves years of sticking objects down one's throat, puking, and dealing with throat sores. This is beyond simply hard work, and all for a craft not merited in conventional mainstream society.

Part of this sacrifice to become a sideshow performer involves a deliberate refusal to participate in the world of "legitimate" work. To guard against that fate, Eric decided to burn his bridges in one of the most brazen ways possible: He tattooed a clown's makeup onto his face, to prevent people from hiring him in case he ever felt the pressure to "sell out." While unfamiliar outsiders might see him as a "lazy freak," Eric perceives himself as motivated. "When you refuse to take a job and you're sleeping on someone's couch, it just puts a strain on everything," he says. "People see you as lazy. I saw myself as determined. I wasn't just lying around on the couch. I was practicing out in the streets, performing every day. I was always being proactive about being a sideshow performer. I'm going to perform. And moreover, I'm going to perform this mostly dead genre." Perseverance eventually led to recognition, and Eric has performed on the televised competition *America's Got Talent* and appeared on the drama series *American Horror Story*. He now makes a living busking in New Orleans, touring the U.S. South in sideshows, and running the Cut Throat Freak Show South circus in venues throughout New Orleans.

For those unconvinced of his work ethic, he also holds a BA in sociology from the University of California–Santa Cruz. He loves sociology; he says it makes him better at everything he does in life.[16] Now, Eric asks me, can I as a sociologist take my ethnographic work to new levels and perform at a freak show or busk on the streets of New Orleans? Yeah, you right.

A SOCIOLOGIST GOES BUSKING

Pantomiming

It's three hours before the New Orleans Saints' first preseason home game of the year, a time in which Tim the Gold Man usually pantomimes on the corner of Decatur and St. Peter streets while taking breaks to drink his cheap forty-ounces-to-freedom beer. With my face painted as an unshaven demented hobo clown—a fleur-de-lis just above both upward curving eyebrows—we walk up Decatur Street before meandering into the Central Business District, heading toward the Superdome.[17] Gold Man carries his trademark gold-painted and (from excessive use and lack of maintenance) deflated mini football, which conveniently also served as a prop with which to ridicule New England Patriot fans during the controversial "Deflategate" scandal.[18]

We reach the Superdome to take turns pantomiming for both locals and tourists. They stare at the spectacle that is Gold Man, fumbling for their cameras and phones to take pictures. Gold Man strikes his usual pose, holding a football with a wide front-to-back stance, feet pointed out and body balanced over the center, like an exaggerated ballet fourth position. It takes talent to remain still in that position. After a few minutes, Gold Man looks at me and, smirking, says, "Well, go ahead man. Do your thing." Feeling awkward, I take the football and position myself as a running back about to stiff-arm a potential defensive player attempting to make a tackle. People almost immediately stop and stare to take pictures; others point fingers in my direction. This continues until a policeman approaches us, explaining that we cannot conduct our business this close to the Superdome. We look for another spot while heading toward Champions Square, a brightly lit area covered in advertisements, where football fans hang out prior to a football game. Although it's a good busking spot, police are intolerant of any disruption to the flow of traffic. We head to Sugar Bowl Drive, near the passageway to the Superdome entrance, where security is stricter than it is at an airport.

After taking turns pantomiming, Gold Man and I collaborate to pantomime together. He'll strike his usual pose, and I'll act as an offensive blocker. The hardest part initially isn't standing entirely motionless for minutes at a time. Rather, it's keeping one's eyes fixed at a specific point and refraining from blinking as much as possible while large groups of people walk past and take pictures. After pantomiming for about an hour it becomes obvious that this work is difficult, and though at times it's entertaining (it's better than an office job, of course), it eventually becomes boring and uncomfortable to stand still for so long counting tips in your head. After we make a couple of dollars, someone hands Gold Man a free ticket for front-row seats, provided he doesn't scalp them. We walk together to watch the Saints, our "local heroes," as the New Orleans sportswriter Peter Finney used to describe them, begin another unpredictable but hopeful season for the diehard "Who Dats."[19] Judging from firsthand experience, pantomiming in the streets of New Orleans, to put it colloquially, ain't for no slouches.

Cut Throat Freak Show South

Today's Cut Throat Freak Show South show at the venue Mags 940 Bar on Elysian Fields Avenue involves risky sideshow circus stunts with its host Eric Odditorium, an assistant clown called Clown Zero, the burlesque clown Jessa Belle, the locally famous Stumps the Clown, and me, a clown-in-training called Cuban Pete the Clown. The show runs on New Orleans time—that is, forty-five minutes late. Things finally get underway at 10:45 p.m. Stumps the Clown plays guitar folk music while dressed as a steampunk hobo, a bottle of peppermint schnapps and orange juice at his side. Following the performance, Eric explains how the "do-it-yourself sideshow" works to the attentive audience, who are drinking beer and screaming such things as "Yessssss, do it," obviously eager to see a bunch of nuts do crazy-dangerous things. On the stage is a pile of broken glass. Stepping on it, Eric asks if everyone can hear the crunching. Satisfied by the response, he continues hosting the event, introducing a mini burlesque show featuring Jessa Belle. Other, more bizarre acts follow. Stumps the Clown gathers audience-provided objects—bags, purses, and the like—and lifts them off the ground with the hooks connected to his earlobes. He follows that by lifting a bowling ball with his ears. Jessa Bell spreads her legs and lifts them in the air, becoming a human version of a carnival ring-toss game. One dollar gets audience members one hula hoop to toss; two dollars gets three hula hoops. Removing one's clothing earns audience members extra hula hoops, one hoop for each article of clothing. One woman wiggles her hips while sliding off her panties. A male audience member pays two dollars to throw three hoops at one time, misses, then takes off his shirt, shorts, underwear, and hat in four more vain attempts to toss the rings over Belle's waiting legs. Another guy allows himself to be blindfolded and takes off his clothes—only to miss with every hoop. Everyone goes outside to smoke and talk during the intermission. It is 11:25 p.m.

The second set starts with a sword-swallowing act, similar to the one Eric performs on Jackson Square, and moves on to "Story Time with Stumps the Clown." Stumps reads quietly to himself while two volunteers from the audience hold his book and support his feet until Jessa Belle, who also happens to be his real-life wife, grabs a hook and pulls him offstage. Jessa then performs another burlesque routine, to the loud shouts of excited audience members.

Eric approaches the microphone in front of stage right next to a large pile of broken vodka bottles to introduce me as a clown-in-training. About two days prior, I had about five minutes of training after drinking heavily much of the night with Eric and his wife, Joe, who both offered some advice on how not to destroy your feet while stomping on glass. I approach the glass, touching it with my hands and carefully scrutinizing the shattered pieces. Eric asks me to place one foot on the glass to start. I approach the pile and hold my naked left foot above it, then step down. What surprises me most is the crunch of the glass as my foot sinks into it. As this happens, I notice the audience watching me perform, not like at some boring academic conference where people fancy themselves saying clever things, but as a sideshow "freak," walking and stomping on glass with all his might. The thought occurs to me that my foot could split wide open. That fear is less daunting than the embarrassment and humiliation such an injury would cause. I place my second foot into the pile and rub it deeper and deeper into the shattered glass, nearly puncturing the skin. Eric orders me, "You need to stomp the glass. Stomp the glass." I look up and ask, "What did the glass do to me?" To which he replies, "Well, nothing yet." I, Cuban Pete the Clown, jump with hesitation but definite determination directly on the glass as Eric reacts, "Ohhhh, geeeez, are you bleeding?" Cuban Pete jumps again, this time a little higher, as Eric gasps along with the audience, "Oh God, that hurts." I jump, more confidently, a final time on the broken glass, then slowly step out of it, carefully wiping off the small pieces stuck to the bottoms of my feet. The feeling of glass crunching under bare feet, slightly piercing the skin, stings with an unusual sharpness but is also almost therapeutic. The crowd applauds as I return to my place with the other stage performers. The weight of fear lifts from my shoulders, and a sense of accomplishment consumes me: I could perform as a glass stomper in New Orleans freak shows should I ever decide to leave academia!

As the final act of the night, Eric performs a second sword-swallowing routine, break-dancing with a sword down his throat while a volunteer from the audience provides a beat-box soundtrack. All the while, Eric's little dog jumps through hoops nearby. Before leaving the stage, Eric lets audience members staple money to his body—money he will keep. For his big finale, he staples his scrotum to his inner thigh for a $75 fee raised by the audience.

But while buskers work the bright urban spaces of the Vieux Carré, street hustlers and con artists exist in the shadows, tricking people out of their cash.

THE HUSTLERS OF NEW ORLEANS: "TWENTY DOLLARS I TELL YOU WHERE YOU GOT DEM SHOES AT"

Two hustlers, one with gold grills in his mouth, approach a white couple, seemingly tourists, holding hands while walking down North Peters Street. One of the hustlers says, hand extended, "Twenty dollars I tell you where you got dem shoes at." The startled white man, thinking this is just a friendly joke, shakes the hand of the hustler. After all, it's rude to not shake hands with another man, and perhaps disrespectful and racist if he is black. The hustler tells him, "You got them shoes on your feet, right here on Decatur Street." Before the hustler even finishes his sentence, he squirts yellowish goo—shoe polish, apparently—onto the young tourist's tennis shoes.

The young man is utterly confused. He lets himself be led to a wall that keeps the couple just outside the view of spectators and, more importantly, cops. These guys use urban space well. While up against the wall, the hustler instructs the tourist to put his foot on the hustler's knee as he wipes it down with a rag—he proceeds to shine this man's tennis shoes. Just as quickly, the other opportunistic hustler squirts goo onto the young white woman's ballet flats and leads her, too, to the wall. It's worth noting that the hustlers never seem to approach single women and only make secondary contact with a male-accompanied woman as part of the hustle directed toward the male victim. Perhaps it's honor among hustlers, reminiscent of Sutherland's honor among thieves, or maybe it's fear of provoking violence from men with egos most certainly offended from having "their women" hustled by another man.[20] Even tourists have a limit to their tolerance.

The men talk to their bewildered clients, informing them that this transaction was twenty dollars per person, ten dollars for guessing correctly and ten dollars for the shine. It is all over as quickly as it began; the man pulls out a twenty-dollar bill and hands it to the man that shined his shoes. The hustler walks away pleased, but the man who shined the woman's shoes informs the man that it was twenty dollars *per person*, and the young man—perhaps not understanding, or perhaps just to get away—tells him that he already paid for his woman and gave the other man the money. This hustler gets upset and says he was supposed to pay them each separately, and the white man continues to explain that he already paid the other guy, that he is sorry. The tourists walk away with fear in their eyes, and the hustler, hoping not to draw outside attention, backs away, looking upset. The couple walks for about a block without saying a word. Meanwhile, the two hustlers get into an argument about the money, one claiming that since he set up the hustle it was his money, the other demanding he get his fair share for playing his part. They conclude their argument as another opportunity appears: another couple far from home. One of the hustlers says, "Twenty dollars I tell you where you got your . . ."—he looks down see that the man wears cheap sandals— "where you got dem *flops* at." He begins to polish the man's flops.

A few minutes later, another group of hustlers approaches three large and plump tourists near Jackson Square as they head to the side entrance of Pat O'Brien's on St. Peter Street. A middle-aged black hustler says to the man, "I like dem shoes. Twenty dollars I tell you where you got dem at," immediately beginning to apply what seems to be shoe polish to his tennis shoes. The hustler continues unapologetically, "Don't feel bad that you've just been had. You ain't the first." I approach the man who just got hustled immediately afterward. He explains to me that the hustler told him, "Twenty dollars I'll tell you where you got them shoes at." One of the women in his group behind him shouts in amusement, "I told him he got them on his feet!" The man, obviously embarrassed but struggling to maintain dignity, admitted he knew it was a hustle, but before he knew it, the man had put the goo on his shoes and "shined" them. He explained that he told the hustler, "Sorry, buddy, all I got's three dollars. Good luck." He theorized that he was targeted because he was a tourist and seemed annoyed at the hustle, but he opened up his wallet and pulled out some money anyway.

These are the hustlers of New Orleans. They might be down and out, but they know how to extract money from those with the cash, time, and privilege to afford travel. They use their creativity and wit to hustle well-educated middle-class tourists out of their money. From nothing, they can get twenty dollars in seconds from a complete stranger. You know where you got dem shoes? They do, more than most.

———————

New Orleans' buskers—the self-made transgressive misfit entrepreneurs of late modernity—find ways to make money while avoiding the low-wage jobs of the city's tourist industry. Judging from my own observations, dozens of interviews, and personal busking experiences, buskers hardly fit the categories of poor, lazy, and undignified. The buskers of New Orleans also enjoy a respect and dignity often unavailable to members of "normal" mainstream society working humdrum conventional jobs.

This chapter reveals how both willfully and socially excluded members of the under- and near-underclasses find creative solutions to make money and find work that grants social status in a postindustrial economy where such prosperity and dignity is becoming harder and harder to come by. Just as the culture of New Orleans showcases its remarkable resilience under heavily oppressive structural conditions, especially in the aftermath of Hurricane Katrina, these creative transgressive misfits of the postindustrial economy, where there is a paucity of good jobs, have created a relatively successful busker culture on the fringes of the tourist economy. This chapter reveals one type of creative response people have to conditions of widespread economic restructuring, which have exacerbated the inequalities between the wealthier and poorer classes. The buskers showcase the refusal of people simply to accept the structural conditions imposed upon them; rather, they find creative and often unique solutions to their collectively experienced problems. The question remains if this showcases human agency or the undying nature of the human spirit to refuse subordination and submission to power and domination. Perhaps it demonstrates both. As inequality increases, we should expect a rise in the number of outsiders refusing to accept the oppressive structural conditions of our times.

Young drummers performing for tips.

Source: Todd Norman.

Busker smoking on Bourbon Street.

Source: Todd Norman.

Tarot-card readers of Jackson Square.

Source: Todd Norman.

Nightlife on Bourbon Street.

Source: Todd Norman.

Marina in New Orleans.

Source: Todd Norman.

Buskers and their buskets.

Source: Todd Norman.

Tremé.

Source: Todd Norman.

Young tap dancers performing for tips.

Source: Todd Norman.

Street life on Pirates Alley in the French Quarter.

Source: Todd Norman.

Brass bands of Jackson Square.

Source: Todd Norman.

Artist on Jackson Square.

Source: Todd Norman.

Gutter punks and their dogs.

Source: Todd Norman.

Poet for hire on Royal Street.

Source: Todd Norman.

Homeless man sleeping on the corner of Decatur and Esplanade.

Source: Todd Norman.

Ladder Man.

Source: Todd Norman.

Gutter punks and their dogs on Decatur.

Source: Todd Norman.

Frenchmen Street food vendor.

Source: Todd Norman.

Night balcony scene overlooking Jackson Square.

Source: Todd Norman.

Brass band performing in Faubourg Marigny at night.

Source: Todd Norman.

A busker, a beggar, and a dog.

Source: Todd Norman.

Traditional brass band second lining.

Source: Todd Norman.

Brass band performing at a memorial in the Bywater.

Source: Todd Norman.

Busker posing on Bourbon Street.

Source: Todd Norman.

Eric Odditorium's Cut Throat Freak Show South.

Source: Todd Norman.

Occultist temple in the Bywater.

Source: Todd Norman.

Marina checking into the homeless shelter.

Source: Todd Norman.

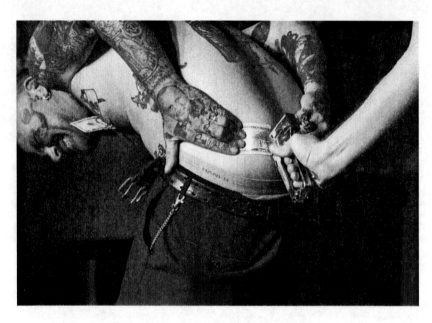

Eric Odditorium gets stapled for money.

Source: Todd Norman.

Pete the Clown stomps on glass.

Source: Todd Norman.

Marina under the Claiborne Overpass.

Source: Todd Norman.

Occult ritual at St. Roch cemetery late at night.

Source: Todd Norman.

Satanic altar in New Orleans.

Source: Todd Norman.

Music-playing buskers in the French Quarter.

Source: Todd Norman.

Young tap-dancing busker.

Source: Todd Norman.

Busker posing for tips.

Source: Todd Norman.

Nocturnal strollers on Bourbon Street.

Source: Todd Norman.

Unkle Loui and Marina shaking hands.

Source: Todd Norman.

Bourbon Street performers.

Source: Todd Norman.

Busker spot during the day and homeless spot at night.

Source: Todd Norman.

Young topless woman poses for tips.

Source: Todd Norman.

Topless busker striking a pose for onlookers.

Source: Todd Norman.

Marina's only possessions upon arrival in New Orleans.
_____
Source: Peter Marina.

Marina bartending on Bourbon Street.
_____
Source: Peter Marina.

Shannon's Poetry While You Wait.

Source: Peter Marina.

Lightfoot and Kesha at Molly's in the Marigny.

Source: Peter Marina.

Shannon the Bourbon Street Whip Girl.

Source: Peter Marina.

THE INFORMAL NOCTURNAL ECONOMY OF FRENCHMEN STREET

Fear of the mob is a superstitious fear. It is based on the idea that there is some mysterious, fundamental difference between rich and poor. . . . But in reality there is no such difference. The mass of the rich and the poor are differentiated by their incomes and nothing else, and the average millionaire is only the average dishwasher dressed in a new suit. Change places, and handy dandy, which is the justice, which is the thief? Everyone who has mixed on equal terms with the poor knows this quite well. But the trouble is that intelligent, cultivated people, the very people who might be expected to have liberal opinions, never do mix with the poor. For what do the majority of educated people know about poverty?

—George Orwell, *Down and Out in Paris and London*

THE MARIGNY STROLL

Steamy midsummer-night dreams flourish in the swampy, romantic, informal nocturnal economy of Frenchmen Street, home to a nightly carnival of decadence centered on dozens of shoulder-to-shoulder music venues. The "Marigny stroll" offers the flâneur an insider's peek into the thick New Orleans nightlife, in the neighborhood where a "new Bourbon Street" emerges, attracting upscale tourists seeking to venture beyond the confines of the Vieux Carré.[1] Frenchmen Street nightlife offers a jolt to the senses of even the most jaded connoisseur of cosmopolitan city life.

The stroll begins at the corner of Decatur Street and Esplanade Avenue, where heavy-metal music spills out of the bar Checkpoint Charlie's and where practicality and pleasure mix at Washateria, a curious establishment in which people can wash their clothes while listening to loud music over beer and burgers. Scores of gutter punks and heavily tattooed derelicts surround the building, smoking cigarettes and joints; others panhandle. Strolling along the lakeside of Decatur Street offers prime sidewalk space for panhandlers and homeless people to sleep, drink, laugh, argue, and beg.[2] About fifteen feet of ambiguous urban real estate sits between the two businesses Checkpoint Charlie's and Vaso Lounge. It has no clear boundaries that separate the public and private rights to space. As a result, gutter punks and panhandling bums—their dogs in tow—use this space to pass the night away with beer and whatever else they can find to drink or smoke.

It is commonly held among street people that police hesitate to arrest beggars and panhandlers with dogs because they have to wait for Animal Control to arrive and take possession of the animal before they can arrest its owner. Numerous police officers in New Orleans confirm that belief. Another tactic of squatters, beggars, panhandlers, and other street folks seeking to avoid run-ins with the law is to utilize ambiguous urban spaces, planting themselves between functioning business establishments to avoid owners who might complain and call the police. Without a clear right to urban sidewalk space, business owners hesitate to call the police, and most police officers generally don't seem to care so long as the sidewalks aren't being blocked.

Vaso Lounge packs huge crowds of tourists and locals, with the music beginning early in the evening and continuing deep into the night. A tattoo joint sits nearby, filled with middle-class kids deciding on designs. The Frenchmen Street scene changes throughout the evening. Formal businesses rule earlier on, but after 11 p.m. dozens of informal street businesses emerge, selling everything from food to jewelry. Business owners complain when vendors sell food on the streets, thus taking away potential customers from their "legitimate" establishments. But it's all fair game once these businesses close at 11 p.m. Like an unspoken second shift, informal vendors reawaken street life as they set up shop the moment managers close their restaurant doors. There are no lulls in the New Orleans night.

Venues such as the Maison, Bamboula's, Blue Nile, d.b.a., 30-90, and the Japanese and tapas bar Yuki Izakaya line Frenchmen Street, playing live music late into the night. This section of the Marigny stroll hosts some of the best music in the New Orleans tourist industry, attracting hundreds of revelers.

Near the closed-down Café Brazil, a brass band shares their souls with the dozens of revelers dancing on the streets, drinks in hand. The horns' cacophony pushes the maximum decibel levels spelled out in the city's noise regulations. The band performs on this corner almost every night, starting at about 10 or 11 p.m.

While strolling past the Three Muses, a man asks, "Y'all need to eat?" He's hoping for some business for his pop-up kitchen. A half-dozen food vendors set up near the Praline Connection, which closes at midnight on weekends and an hour or two earlier on weekdays, cattycorner from the tourist-packed Dat Dog, which sells overstuffed and overindulgent—but also overpriced—hot dogs and sausages. Many of these vendors cook and sell food on the streets without licenses but manage to avoid city authorities, who only occasionally ticket illegal vendors for their transgressions.

During restaurant hours, dozens of vendors pack Frenchmen Street to sell necklaces, jewelry, oils, soaps, and other artistic creations and crafts. This scene too transforms when the restaurants close, between 10 p.m. and midnight: Many of the arts and crafts vendors pack up their products, making room for the street-food vendors. Food vendors use pushcarts, charcoal barbeque grills on trucks, and other various four-wheeled mobile carts to cook and sell their culinary delights. Some vendors cook on open fires on the side of the sidewalk; one strange mobile vending apparatus called Hoodoo Barbecue is an actual barbecue pit on wheels. These guys cook and sell ribs, jerk chicken, and what they call a "ghetto burger" for $10, "hot diggity dogs" for $7, and "I got some of dat New Orleans sausage" for $8. Some of these vendors even accept credit cards for a $2 surcharge. A man with a big white ice chest sells canned Abita Beer and other popular brands for $3.

Beggars, homeless travelers, and gutter punks squat on the corner of Frenchmen Street and Chartres near the entrance to Faubourg Marigny Art and Books, which closes each night just as the nocturnal economy is gearing up. One holds a sign stating, "Too broke too travel too ugly too prostitute"; another makes jewelry. One gutter punk in the crowd of young homeless people yells to a female stroller, "You are beautiful. Will you take me and my wife home with you? We promise not to bite, unless you want us to." Another face-tattooed man of Roma descent says, "Excuse me sir, but can you spare a smile?"

On Frenchmen Street beyond Chartres, Café Negril, which once played mainly Jamaican music, now caters to local sounds. A man in front of Café Negril sells pepperoni, bacon, and veggie pizzas. Some food vendors make deals with the business owners to operate their informal businesses inside formal ones in order to obtain a stationary license. This pop-up food vendor outside of Café Negril has obviously struck a profitable deal.

The Apple Barrel down the road plays mainly live rock-and-roll and honky-tonk music while locals and tourists wait to eat in the restaurant above, Adolfo's seafood and Creole Italian restaurant. Crowds of people gather between the Apple Barrel and the Frenchman Market, where dozens of licensed vendors display their art, jewelry, and crafts. A woman with an unhappy cat on a leash allows people to take pictures of her for a buck. Other artists just outside the formal Frenchman Art Market sell their paintings, making use of the cracks and crevices of informal urban spaces.

The famous 1920s-jazz-themed music venue Spotted Cat next to the French Art Market is a magnet for swirling and twirling tourists swinging to the beautiful voices of Meschiya Lake of the Little Big Horns or the funk sounds of the New Orleans Cottonmouth Kings and Washboard Chaz Trio. The more upscale music venue Snug Harbor puts on a more controlled New York–style jazz; its audience waits for the show to start in the club's dimly lit bar. Meanwhile, tourists at the nearby Marigny Brasserie on Royal Street discuss everything from politics to philosophy while sipping cocktails on the sidewalk outside.

The faint sounds of clacking typewriters in the Crescent City's night air can be heard from the Frenchmen Street sidewalk poets, who sell poetry on commission to Marigny strollers. A barefoot young woman named Shannon writes poems for tourists while thinking about a world she longs to discover fully on this Frenchmen Street of broken dreams.

THE FRENCHMEN STREET POETS

The smartly dressed poets of Frenchmen Street set up informal writing spaces in front of Bicycle Michael's bike shop. They enhance the charm of Frenchmen Street, a "classier" version of Bourbon Street, where the sounds of the "real" New Orleans still live, even if altered to cater to the palettes of outsiders. Shannon used to be a Bourbon Street "whip girl": she engaged in light sadomasochism for a fee by whipping "naughty" tourists for a buck. Now she writes poetry on commission in the much safer environment of Frenchmen Street. She ran away from home at the age of fourteen, leaving behind her "crack-addicted mother" and a verbally abusive father, and seems in her element on Frenchmen Street, writing poetry for curious and inquisitive tourists.

Shannon's life experiences shaped a dark inner strength that lurks beneath her genteel demeanor and cat-with-the-canary smile. She's been on her own through the majority of her teenage years, hopping trains, squatting in abandoned houses, dumpster diving, and making a living in the informal street economy. Besides whipping people on Bourbon Street in the past, she used to blow fire on Frenchmen Street, until the cops cracked down on the practice. She earns enough writing street poetry to pay for a room in an apartment in the St. Roch neighborhood and for essential expenses. Shannon sells each poem for about $20.[3] She earns about $100 on slow nights. On busier nights, especially during tourist season and festivals, she makes between $200 and $300 a night. Working four or five nights a week, she can make somewhere between $400 and $1,500 of untaxed income weekly—more money than the meager wages of many full- and part-time New Orleans service-industry workers. It becomes obvious sitting next to Shannon for a few hours that she does good business, with tourists sometimes lining up for a poem.

Selling poems for money on the streets without a vendor's license skirts the lines of legality, but by doing so Frenchmen Street poets like Shannon are able to avoid working conventional jobs. It's also a relatively safe environment in which to make money on the streets, though it does bring certain aggravations. Some tourists harass them, trying to get them to compete to find out who is the better poet. One of the poets explained to a customer that they are all part of a poet's collective and work in cooperation. Whenever tourists get out of hand or become verbally abusive, Shannon scolds them, which has earned her the nickname the "Terror of Frenchmen Street."

POETRY WHILE YOU WAIT

A few of us sit talking about life and poetry, watching tourists stroll by. Shannon's friend Shane wants to try writing poetry on Frenchmen Street for the first time; I spend most of my time talking with Shannon. A man and woman, who appear to be in their mid-thirties, approach and ask how the arrangement works. Shannon responds, "Tell me a topic for a poem and I'll write it." They think about a topic with a grin while asking about the price. "It's whatever you want to pay. Most people give $20 but whatever is fine," she says. The two lovers talk privately, then return.

> *Male customer:* We would like poetry while we wait.
> *Shannon:* We're a team effort tonight! So you're gonna get a handcrafted poem with two poets' work into it. Where y'all from?
> *Female customer:* Missouri.
> *Shannon:* I like Missouri. People are good people in Missouri. I got stuck in Missouri in the pouring-down rain hitchhiking, and people were still picking me up. Good people in Missouri.
> *Male customer:* Absinthe.

Shannon nods and immediately begins typing.

The couple stands back a few feet talking, keeping one eye on Shannon and the other on the vibrant street life of Frenchmen Street. They give off a certain air of "prove it," as if highly suspect of the ability of these street writers to produce a decent poem, especially on the topic chosen. Shannon types first before giving the inexperienced and somewhat nervous Shane the opportunity to contribute. As Shane struggles to find words, Shannon and I discuss her life making money on the streets:

> *Marina:* So how long have you been doing this?
> *Shannon:* I've been writing I guess since I was fourteen.

Marina: And what about working on the streets?

Shannon: I've been making money on the streets since I first got on the
road, which is when I was about to turn twenty-one.

Marina: Do you have an itch to travel?

Shannon: I've hitchhiked for four years. I'm done. I rode my first freight
train, so I'm gonna go on a freight train again soon. . . . [*to Shane*] I can
help you with that if you want . . . [*to Marina*] I got a *Crew Change*
[*Guide*], so I'll definitely be back traveling for the freight train soon.
Sooner than later, but yeah.

Marina: How often do you work here?

Shannon: Depends on how much money I make for the night. Yeah, I'm
just grateful I don't have a nine-to-five.

Marina: Well, yeah, I wouldn't want a nine-to-five either.

Shannon: And I do know the other poets. We all get along really swimmingly
well. And I feel really good about that. We're not competing.

Shannon returns to the typewriter and, after a few minutes, approaches the
couple to read the poem, first offering a preface that explains the ideas behind
the words. She explains, "When you said 'absinthe' I totally thought about
what the French call an orgasm, a *petit mort*, like a mini death. And I think
that's where we took this." Shane and Shannon read the poem out load dramatically, open-mic night / poetry jam session style.

Absinthe

Le absent
A petite mort of this vision
I no longer wish to see.
Instead—courage is rarely even
Often it is borrowed
As time is spent borrowing its
Seasons from the sun
No time is changeable
Only equal to its own death
Precursor: death is a lion—
Be cautious of its teeth

As you see the smiling fangs beginning
To surface over the red oak of the bar.
The way the lady flickers her wrist
To pour the decanter over the sugar cube
Romantically, as if to say . . .
Are you sure you're ready
       for wormwood
         today?

Shannon and Shane 5/29/14 New Orleans

The surprised couple seems impressed with the poem, with the male stating, "Ah, that's fantastic. That was *way* better than I thought you guys were gonna do."

Shannon responds with a smile and twinkle in her hazel eyes. "Thanks, honey. I've got an envelope for you, too." She places the poem in an envelope. The pleased couple hand her a twenty-dollar bill as she hands them their poem.

THE INFORMAL WORK SHIFT OF
THE FRENCHMEN STREET POET

Shannon sets up her table with a vintage manual typewriter, giving it the air of the romantic poetry of 1950s beatnik America. She puts a rug under the table and adds a cushion to sit on. She sits with her legs folded on the cushion without a chair to type. Her sign, "Poetry While You Wait!" rests on the front side of the table. Mobs of people pass her on the sidewalk throughout the night, often taking notice of the poets. A high-school teacher asks for a poem titled "Education" for her students back at home; a voluptuously rotund couple fancies a poem about "Long-Distance Relationships." Another nondescript guy chooses "the middle of the Earth." Some poets ask for more insight into the topic or inquire more about the person requesting the poem. Other poets take on the topic immediately without any questions. The poet taps away at the typewriter while the customer waits. The writing usually takes less than five minutes. Once done, the poet stands up and reads the customer the poem. The poet places the poem into an envelope, and a carbon-copy goes into the writer's personal files.

The Frenchmen Street poets want to do more than simply write a poem. They want to impress their customers with something memorable. It's a matter of pride. They put their artistic talents and skills on display for public scrutiny. Offering a product that brings delight to customers brings a feeling of work satisfaction, a sense of fulfillment often not found in many conventional jobs. A young man in a group of five tourists asks for a poem about "Surfaces and Surfaces." After typing the poem, Shannon reads it to the customer with a sense of passion:

Surfaces and Surfaces

Glimmering between the spectrum of my untrained eyes
Struggling to find anything deeper than the basic ties
Hiding behind what is already in sight
Irony is deaf to fire,
Bring light to the breadth of the living edge.
Nothing is wasteful, nothing is gone.
It is all here,
Breathing just above below between the surface,
Edging away from the philosophy into the apple of time.
We bite into it as though it is the last taste we will ever have of life.

The male customer, grinning widely, exclaims, "Wow, thank you. This is incredible."

I think of Weber's charismatic figure, who gains authoritative legitimacy through personal charm. It's not important if these poets actually have the talent and insight of a "real" poet; they must only be able to convince others that they do. Shannon possesses the qualities of a charismatic street poet. She often writes her own poems, like one on gentrification, simply for love of writing and the pursuit of creativity.

Poetry Gentrification

There I was, Big Easy Christmas . . .
Pauline Street in the Bywater.
I was strolling out in the morning . . . you know

Around 12:30.
Grabbing my coffee
Getting my morning pack of cigarettes.
There he was . . . Blonde Santa.
I could even see the hair, peeking out from—
Under the glossy white wig.
"But Santa is supposed to be black, in this neighborhood."
What is happening?
Before I even got to give my usual clerk
My cash and ingest my morning sanity . . .
I'm faced with the reality that yuppies have invaded our safe haven.
We created it—
now it's being taken away ripped from our hands
ON THIS DAY JESUS WAS BORN!
What the fuck
Do I have to blast Mr. Bungle at 3 AM to make you fuckers leave my part
    of town?
Or will I just have to watch you infest us
Then my home and life here will slip out under me.
Tax crisis . . . Ugh. Maybe I'll try the seventh ward.

One gets the feeling she's on her way toward becoming a "real" poet.

FESTIVALS AND THE INFORMAL STREET ECONOMY

Thousands of tourists flock to New Orleans during its many music festivals, clogging the limited urban space in the compact French Quarter and Faubourg Marigny. The Essence Music Festival, which is held every Fourth of July weekend, celebrates black culture and music with funk, gospel, jazz, blues, R&B, soul, reggae, and hip-hop concerts, among other ancillary events.

These festivals attract thousands of locals and tourists, but they also make the police irritable. According to some police officers of the Eighth District, or the Vieux Carré, cops typically ignore the food vendors and poetry writers on Frenchmen Street unless someone calls the police. This attitude changes during large festivals, when the police enforce laws they otherwise largely ignore.

Looking at the Code of Ordinances for the city government of New Orleans, street poets seem to fall best under codes referring to rules of conduct for artists.[4] This city ordinance requires artists to hold a permit in order to conduct business or entertainment on public spaces. The following ordinance sections show the city requirements on street artists as well as on those who use public space to conduct business.

DIVISION 3. ARTISTS

Sec. 110-122. Permits required.

(a) Every person who shall desire to use the public streets, sidewalks or public places or private places of business establishments for the conduct of any of the businesses or callings hereinafter set forth or to hold meetings or rallies or public entertainment in private or public halls or places shall first apply to and obtain from the department of finance a permit.

Sec. 30-1453. License required.

Any person wishing to engage in business as a street entertainer shall first obtain a license under this chapter.

(Code 1956, § 46-1)

Sec. 30-1455. Fee.

The fee for a license under this article shall be $15.00 per year.

(Code 1956, § 46-1(18))

New Orleans city ordinances also forbid blocking the sidewalk or obstructing public passages. This can easily, at the officer's discretion, be applied to poets accepting commissions, even if through donation, for their artistic labors. The ordinance reads as follows:

Sec. 54–401.—Obstructing public passages.

(a) No person shall willfully obstruct the free, convenient and normal use of any public sidewalk, street, highway, bridge, alley, road, or other passageway, or the entrance, corridor or passage of any public building, structure, watercraft or ferry, by impeding, hindering, stifling, retarding or restraining traffic or passage thereon or therein.

Violations can result in a misdemeanor and a $500 fine or six months in jail. Unlike the easier-to-obtain busker licenses in Jackson Square, access to street-vendor licenses is much more difficult.

City Hall is a complex bureaucratic labyrinth of dozens of departments and agencies that control everything from recreation and culture to politics and economics. The bureaucracy loves confusion and ambiguity, for it retains the power to interpret the meanings of all that is blurry. Science fails us in our attempt to enter this world and understand it. Turning to voodoo may prove more effective.[5]

VOODOO BUREAUCRACY IN NEW ORLEANS CITY HALL

Getting an occupational license means entering into a great hall of corruption and scandal—New Orleans City Hall. After paying for a parking permit, entering the disenchanted domains of City Hall requires undergoing airport-level security checks, including emptying pockets and walking through an old-school TSA body scanner.

The Bureau of Revenue is located on the first floor, in room 1W15. It's a rather plain white room with several customer-service windows on both sides and many cubicles and empty office desks. There's a flat-screen television with a list of names and departments as well as a sign-in sheet on the front desk with six names on it. Within ten minutes of entering, I am told to wait in another room with about a dozen other folks with sad faces. Citizens wait while sitting in rows of chairs sandwiched between the windows, not all of which have clerks behind them. One man sits at a desk filling out a form. Another man sits slumped in his chair while talking quietly to a female clerk at a window.

After a few minutes, a tall, informally dressed man approaches me. I request information about getting permission to sell poetry, food, and other material items on the street. He tells me to wait while he disappears into a back room, reappearing shortly with multiple applications for various permits and licenses. I ask several questions about what street activities require permits and licenses and where I can get them. He struggles to answer the questions with any degree of clarity. Instead, he shuffles through the papers, skimming them as we talk. Without much information, I turn to leave, only to hear a man sitting outside one of the customer-service windows attempt to answer my question. He explains that in order to get a permit to play music and to sell poetry and food in the French Quarter, one needs to go upstairs to the Vieux Carré Commission (VCC) on the seventh floor, room 7E05. The woman behind the glass confirms this.

The entire seventh floor serves the singular purpose of obtaining permits and licenses in the city of New Orleans. A woman sitting at a desk ignores my request for information. Instead, she asks my name, types it into her computer, and tells me to have a seat, explaining that her supervisor will help me. Behind her, a weary-looking security guard sits next to the receptionist's desk. The people in the room wait for separate services from various departments: Driver/Operator, Building Permit, City Planner on Duty, Occupational License, Payment, Special Event, Building Inspections. A large TV screen displays a list of names next to the department that will (hopefully) serve them. My name pops up in the Occupational License category. The names called follow no obvious order. During my hour-long wait, about a half-dozen people who entered the waiting room after me receive service first. What kind of bureaucracy is this? Voodoo bureaucracy.

The screen further details the people currently being served, and next to each name and corresponding department, the proper window, room, or office cubicle. When my name is finally called, a large woman with a kind face greets me, asking how she can help. I first inquire about obtaining a license for selling food on Frenchmen Street.

Marina: I'm curious about getting either a mobile or stationary cart and finding out what's the difference between the two. I want a license to sell food on Frenchmen Street.

Clerk: That's prohibited. Is it the Frenchmen part that's in the French Quarter or outside?

Marina: [Realizing that she does not know New Orleans geography well, given that no part of Frenchmen Street is in the Quarter] In the Marigny.

Clerk: All right. Well, the food cart has to be a truck.

Marina: Okay. I see people with tables outside selling food, no?

Clerk: You're not supposed to do that.

Marina: So you can't get a permit for that?

Clerk: No.

Marina: So you have to have an actual truck with wheels on it?

Clerk: Yeah. And these [*points to application*] are the guidelines in order to have a food truck. So the truck has to be able to drive itself. It can't be pushed or pulled. It also tells you to, in the back, cause I don't, I'm not familiar with all the rules, but I know that, I believe, in the CBD and the Vieux Carré, you're prohibited from having a food truck and selling, especially around here.

Marina: [*Confused*] What?

Clerk: Unless you're franchised. And in order to be franchised, you have to go through the actual food-truck process first, and then you can go through city council.

Marina: Okay.

Clerk: Which is a headache. But if you just read all of this, it'll tell you, I think, your rules, your dos and don'ts, but this is what we would need from you: vehicle registration, the automobile insurance, liability insurance, health department and fire prevention approval, a photo of the truck, and a copy of the driver's license for each driver that's gonna drive the truck.

Remembering that a dozen food vendors cook hamburgers, hotdogs, chicken, and sausage right on Frenchmen Street, sometimes with open fires, I wonder about the legality of street cooking. Many of the street vendors told me that they had licenses, but it was impossible to know if this was accurate with any degree of certainty. The woman tells me that cooking on the streets is illegal unless the Health Department approves it. Vendors can receive permission to cook within—and sell food from—enclosed spaces in vehicles that can drive away after a four-hour time limit.

The conversation switches to selling jewelry.

Marina: What if I want to sell jewelry, to have a table with my own jewelry setup?

Clerk: You would be a mobile vendor. You would go straight downstairs to Revenue and see if they'll let you do it.

Marina: On the first floor? I just came from the first floor, they told me to go to the seventh floor.

Clerk: Yeah. Chances are they're not gonna do it unless you're already set up as a vendor.

Marina: Oh, unless you're already set up? So they don't allow newcomers to come in?

Clerk: Um, if you're not properly licensed. That's why I said if they let you do it. Is it gonna be on the street or sidewalk?

Marina: Sidewalk, on Frenchmen Street.

Clerk: So it's on the sidewalk. Usually they don't let you sit on the sidewalk, you're not supposed to, unless you get permission. And it wouldn't be from us; it would be directly from Revenue on the first floor. Let 'em know it's for mobile vending. So they won't send you up here.

Apparently, according to the clerk, the seventh floor issues, for a fee of course, licenses for mobile food trucks that must be "self-propelled" in order to receive permission to sell food. The first floor, she explains, issues licenses for stationary, nonfood vendors. We begin to discuss the different types of carts used for mobile and stationary vending services.

Clerk: Pushcarts. Pushcarts and I think animal drawn. I'm having a brain fart. I can't remember.

Marina: Animal drawn? What kind of animals? There are so many licenses for so many things.

Clerk: It's too many. "Push carts and animal-drawn food vending." I'm gonna print this out for you, too.

As it turns out, pushcarts involve only food, and selling other items such as jewelry involves stationary carts. While obtaining a stationary license involves dealing with the first floor—again—she tells me that setting up in the Vieux Carré District requires obtaining permission from the French Market Corporation, a public-benefits corporation that oversees the French Quarter's famous open-air French Market (but not the entire French Quarter). But she adds, "If it's already an established market, then you would go directly through Revenue, you know, to sell the jewelry." It remains unclear what exactly is an "established market" and how to become "established." It's also unclear where to get licenses for different types of vending and different types of products in different parts of the city. What's happening here? This is some strange magic being conjured in this voodoo bureaucracy.

> *Clerk:* If you were going to be selling directly, selling your own stuff, it wouldn't be us. It would be directed to Revenue. Yep, mmhmm. I am losing it; this is a food cart. Read over that. While you read, I'm going to read it too, because I don't do that many food carts. I usually do just the trucks.
>
> *Marina:* Just the trucks, okay. And the food carts is just when you have to push it yourself.
>
> *Clerk:* Yes. Those are the only ones you can push. I'm trying to figure out where you can push it at.

Perplexed, the clerk shuffles the papers. None of the forms seems to be the right one. She turns from the stacks of forms strewn across her desk to the slow-running computer to look for more information. She seems almost comfortable with the confusion. She is clearly used to it. The conversation continues after two or three minutes of sifting through materials and Internet sites.

> *Marina:* It says, "For vendors that were operating legally in the CBD, the Vieux Carré, or residential prior to the adoption of this ordinance . . ."
>
> *Clerk:* That's the one that already had the right to sell in this area and in the French Quarter. Now they have prohibited new applications. And I think you have to get special permission, or they're just saying no.
>
> *Marina:* Okay. How long might they do that for?
>
> *Clerk:* Um, I don't know. [*She appears confused. Apparently, this voodoo bureaucracy works both ways.*]
>
> *Marina:* So that means I can't have a push cart right now then?
>
> *Clerk:* [*Shuffling through more papers and forms*] Not in that area. You can probably do it in any other areas that are allowed, because I think online they have a map of where you can go, it just can't be the French Quarter or the CBD area, certain CBD areas. [*Looking at computer screen while clicking on various sites within the city's search engine.*] This doesn't really say what you can and can't do with a pushcart.

Other than that it has to be exclusively powered by a horse, donkey, mule, or a bicycle.

Pushcart vendors must be completely mobile and without permanent fixed locations. The difference between a pushcart and a mobile food truck is that the pushcart can sell prepackaged foods like hot dogs; mobile food trucks can sell fresh produce and raw meat and seafood. Mobile food trucks can use fires to cook within their enclosed trucks; pushcart vendors are denied the right to cook on the streets using open fires.

We finally get to the issue of writing poetry on commission.

> *Marina:* I'm thinking about writing poetry on commission on Frenchmen Street with a little typewriter and table. Do I need a permit just to write poems for tourists who might approach me saying, "Hey I'd like to have a poem!" And I write 'em a poem and hand it to them. Do I need a permit?

The clerk tilts her head in utter confusion. Her eyes glance toward her desk and, as if by instinct, begins to shuffle through more papers and forms, looking for anything to use that might satisfy my question. She finally admits:

> *Clerk:* I don't know. I haven't heard of anything like that before, but I don't think so.
>
> *Marina:* So would the police ever come up to me?
>
> *Clerk:* They probably will. Yeah, because they do it to everybody. They're gonna probably ask if you have a license to do what you're doing if you're making a profit. That's the only thing. If you're collecting money, they're going to uh . . . they'll probably ask you if you have an occupational license. If you want to, you can check with Revenue, stop there to see if you need one. And if you do, then they will be the ones to issue it because it would fall under mobile vending.
>
> *Marina:* Why would it fall under mobile vending?
>
> *Clerk:* I could be wrong, and I hate to send you down there 'cause it's like opening a can of worms.

A can of worms? How so? She explains that if Revenue agents become aware of writers making money from selling poetry, they will go out searching for them.

> *Marina:* What do you mean?
>
> *Clerk:* They may look for you, 'cause they have Revenue agents that canvas the city.
>
> *Marina:* Revenue agents? They go around looking for people to take their money? That sounds like hired hit men.
>
> *Clerk:* Yeah. Revenue is on the job. A lot of people don't know that. That's how a lot of people get busted. 'Cause they have Revenue agents and they go and check businesses on a regular basis to see if they paid their taxes, if they're doing illegal sales, everything, so be careful. That's why I almost hate to send you down there.
>
> *Marina:* Wow. It's modern-day Storyville, like where the police collected the taxes in front of brothels.
>
> *Clerk:* Yeah, exactly. But just be careful. If nobody stops you or comes up to you and asks you anything, you should be okay. I don't see why it wouldn't be. But there's a chance that the police will ask you if you have an occupational license. And they do have the right to stop you if you don't. It might be a "quality-of-life" police officer.

It becomes obvious that going to the first floor to inquire about getting a license to write poetry might unintentionally capture the attention of Revenue agents at the expense of the Frenchmen Street poets. Instead, I ask about the potential punishments for writing poetry without City Hall's permission. While Revenue agents may require writers to obtain a license, the police can give citations and require that they obtain a license.

The City Hall clerk offers advice to avoid trouble with ordinance enforcers. This involves making side deals with business owners to use their private spaces to conduct business. This is exactly how some vendors operate, like the man selling pizza in front of Café Negril. Other people, like the Frenchmen Street poets, receive informal approval from businesses like Bicycle Michael's that are closed during the evening hours when the poets write. In short, the clerk recommends exploiting loopholes to avoid dealing with the legal corporation she represents. Extraordinarily, she also admits that it is better to risk problems with citations from the police and trouble with Revenue agents than to attempt to obtain a legal license to write poetry on commission to tourists. My goal now is to go directly to the Eighth District police station to ask the police their thoughts on writing poetry on Frenchmen Street.

THE EIGHTH DISTRICT "VIEUX CARRÉ" POLICE STATION

My return to Royal Street and the 196-year-old building that houses the New Orleans Police Department's Eighth District station triggers a warm rush of nostalgia. These are the old stomping grounds of my father, a retired NOPD cop and former lieutenant of the Eighth District, also known as the Vieux Carré District. As a high-school kid in the 1990s, many were the times I would hop on the St. Charles Avenue streetcar in front of De La Salle High School and ride down here to meet Pop. He was my hero. I was in awe.

This time is different, though. Now when I walk through the front doors of the station house—originally built to house the Bank of Louisiana and still retaining hints of its former stately life—I don't do so as an adoring son. I am a regular citizen requesting information from annoyed and disinterested police officers about making money on the streets. It seems strange and unfamiliar. As luck would have it, the day I visit the station, and with the heat index exceeding 100 degrees, the New Orleans Sewage and Water Board has issued a temporary Boil Water Advisory, stating that customers on the city's east bank—where the Eighth District is located—are "advised not to drink, make ice, brush teeth, bathe or shower, prepare or rinse food with tap water unless it has been properly disinfected, until further notice."[6] Perhaps the heat and lack of running water will make the police even friendlier than usual.

The well-air-conditioned Vieux Carré police station includes a small museum displaying various police artifacts, from old-school billy clubs to classic guns. A young female police officer sits to the left of a large male cop in uniform. I tell him about my plans to write poetry on commission on Frenchmen Street for tourists and ask if the police would give me any trouble for such activities. The police department, he says, has bigger problems to deal with than unlicensed poets and that I shouldn't expect any trouble from them.

He goes on to explain the well-known fact that people always play music down on Frenchmen Street and that it's usually no problem. This is New Orleans after all. I ask about selling food, and he says that the only time the police have any trouble with food vendors on the street is when they try to sell in front of restaurants and the restaurants call the police to complain. Even then, he says, the cops usually first tell the vendor to move along, maybe issuing a citation depending on how the encounter goes. Police officers don't usually bother going to Frenchmen Street to look for trouble with people selling things. There are too many other problems to worry about. He admits that this is always susceptible to a police officer's discretion, but he otherwise encourages me to write poetry on the streets for commission. Go for it, he says. No big deal.

FROM FRENCHMEN POET TO BOURBON STREET WHIP GIRL

The combination of the city's legal system, a major music festival, and police officers' "discretion" led to a series of events that doomed, at least temporarily, Shannon's Frenchmen Street poet days, returning her to a life of playing the risky role of Bourbon Street whip girl before taking off to hop trains throughout the United States.

Just days before Essence Fest, while squatting in an abandoned, Hurricane Katrina–torn house on the outskirts of the St. Roch neighborhood, Shannon and I talked further about the hazards of life as a street performer in New Orleans. She discussed the epidemic of violence that mars the city streets, the aggressive street panhandlers, and the drunken strollers on Frenchmen Street, who are becoming more and more like Bourbon Street tourists. She talked about the problems that cops sometimes give to various street performers during certain times of the year and the fluctuating winds of the city's political climate.

God. I'm afraid for my future as a street performer. I'm terrified. This is all I want to do. I have so much to offer people. When I did whippings, I would take three hours to get ready. I put my heart into this, into this costume, into this character and I went out there and I bent over people with my heart and I whipped them and they got amazing pictures and the best fucking story to tell everybody at home about Whip Girl. That's what I was called, Whip Girl. Just like poetry, you know? I'm hand-crafting masterpieces and throwing them off to people.

She explained how whipping drunk and often belligerent tourists on Bourbon Street, while satisfying and stimulating all at once, now feels too dangerous, especially as the nightlife and heavy drinking continues on into the late hours. Men often paw at her practically naked breasts, which are covered only with pasties, as well as attempting to grab her ass through her short-shorts. Writing poetry for tourists on Frenchmen Street offered a more creative and stable—and definitely safer—place to make money on the streets. Any outsider could sense the tranquil mood of the street writers using their creative talents to eke out a living while attempting to make careers as budding poets. But this summer's tourist season has been slow, and that makes writing poetry on commission difficult. As the number of tourists dwindles, so does the amount of money pouring into the city and into the nearly empty pockets of the Frenchmen Street poets. Shannon just wants to make enough money to hop freight trains and hitchhike to New Jersey to see her grandmother. All she needs is a few more dollars, a few days of poetry writing to fund her travels.

On Thursday, July 2, a New Orleans police officer approached Shannon while she was writing poetry on Frenchmen Street. This was at the beginning of Essence Fest, and the cops tend to enforce city ordinances more strictly during large festivals. The cop ordered her to stop writing poetry, threatening to issue her a summons or ticket if she refused to comply. As a result, the next day, Shannon transformed from the tranquil Frenchmen Street poet back into the Bourbon Street whip girl. Of course, the irony is that poets get rousted by the cops but whip girls are allowed to operate. This raises the question as to why would cops favor one over the other and, further, why they would favor a seemingly less respectable street trade over the more respectable one. Shannon pulls out her leather whip and wooden spanking board with the words "Whippings a buck" and goes wild taking her frustration out on tourists.

Visibly angry, Shannon tells almost her entire life story while sitting without her typewriter in front of Bicycle Michael's, now sadly devoid of the usual poets. But Shannon reclaims this urban space as the place where she blasts out the events that led her to run away from home.

> I was abused by my father in many ways that went deeper than just him hitting me. It was like the control and the shattering of my self-image. He never sexually abused me, but he had an emotional reign of terror and manipulation on me and my family, and my mother was a prostitute and a crack addict. And my father always told me that I was going to wind up like my mother. And when he hit me too hard, and I went to school and I said I'm not going back there, I'm going to run away and you're never going to see me again. I got put into the [custody of the] state of New Jersey. Very shortly after, I discovered drugs and alcohol. So I used them. . . . Then I finished out eighth grade and moved with my mother after being in foster care and group homes, a lot of group homes, a lot of foster care, and then my mom got on her shit [drugs], then I went to live with my mom, because it was like free rein, because I could sneak out of the house. My mom was doing drugs, so I used it as a manipulation tool to get what I wanted. I stole her drugs. I stole her money for drugs. It was a really easy place for me to be a master manipulator because my father taught me to be a master manipulator. And I shed all of these things. But I basically used the drugs to hide from myself for many years, and then I stopped doing drugs when I was seventeen, and then I started traveling when I was twenty.

As she speaks, a couple approaches us, looking for marijuana. They immediately notice Shannon and her nearly exposed breasts. They want to know what she does with that whip and spanking board. Shannon is more than happy to oblige. The once calm and peaceful Frenchmen poet transforms as she vents her anger on the drunken, preppy couple. She spews pure rage that seems beyond her control and that transforms her into a real terror of Frenchmen Street.

Shannon whipped the male for nearly an hour in front of his girlfriend, who displayed mixed emotions, a likely result of drunken sexual play and jealousy. Eventually Shannon and the girlfriend took turns whipping him. The scene finally ended with Shannon ridiculing the couple and ordering them to leave. They did not offer to pay her for her services.

The couple fades into the distance as Shannon turns to me, saying in frustration, "This is what I do, and I'm getting out of here. Fuck those police wanting to give me a ticket. I'm jumping on trains again. I'll make money whipping tourists on Bourbon Street tomorrow and I'm out of here the next day. This is my life story; this is my life." The next day Shannon became the Bourbon Street whip girl one final time before jumping on the train with her "hobo gold."[7] Before she left, she gave me her typewriter so I too could write poetry on Frenchmen Street in my days down and out in New Orleans.

———————

Poets and artists find innovative ways to make a living on the streets of New Orleans. They may skirt legality, but they contribute to the vibrant Frenchmen Street life. As Jane Jacobs would say, as with Duneier's sidewalk vendors, these poets, artists, and vendors increase the eyes on the streets in a dangerous city. Thanks to them, Frenchmen Street remains one of the safest places in the city for locals and tourists; you can walk around as if in a protective cloud. Poets like Shannon are part of the atmosphere of Frenchmen Street, along with the flickering French Quarter gas lanterns, jazz-funk brass bands, the scents of spicy seafood gumbo, the thick humidity, bright moonlight, and the smell of the mighty Mississippi, all contributing to the city's intoxicating and romantic aura, which promises that magic is somehow still possible in the cold, sterile, disenchanted world of modern life. But police "discretion" took that magic away.

Police discretion makes the normally unenforced rules against the informal Frenchmen Street workers seem arbitrary and unjust. While it's clear that enforcing city laws against poets, vendors, and brass bands would be punitive and against the economic interests of the tourist industry, the informal and arbitrary enforcement of the rules creates a sense of uncertainty and insecurity. It also threatens the financial security of these informal workers and heightens the levels of distrust against the police. This type of police discretion drove Shannon away, and it threatens many others who perform on the streets of this great musical city. Music and other art forms in New Orleans express the soul—and a city without soul loses its breath. New Orleans police officers and the Faubourg Marigny residents should use better discretion when considering the soul of the city in which they live.

CHAPTER 6

CITY SQUATTING AND URBAN CAMPING

It is curious how people take it for granted that they have a right to preach at you and pray over you as soon as your income falls below a certain level.

—George Orwell, *Down and Out in Paris and London*

THE JOURNEY TO URBAN CAMPING

Four of us sit outside Flora's Coffee Shop and Gallery on the corner of Franklin Avenue and Rue Royal in Faubourg Marigny, waiting for the New Orleans rain gods to get bored. Earlier, Shannon had texted me to "bring a tarp, a sheet, bug net, water, food for morning, and a flashlight or headlamp." She's wary of the rain and the trouble it might bring because, after a life of traveling, squatting, and hopping freight trains or grainer trains, her feet are sensitive to long periods of being wet. We decide to throw caution to the wind and scope out a place to squat for the night. Our urban camping adventure begins.

It is about 4 p.m. as we mount our bikes and point north toward St. Claude. We head over the train tracks and pedal east toward the Ninth Ward on a road in such a state of decay that it is more potholes than paved street. We pass abandoned, graffiti-covered buildings and a vacant lot containing filthy mattresses, sofas, and tires. Even the trees look derelict. To the left, a partially boarded-up warehouse has graffiti announcing "Vampire Sex," "Oye," and "Finish Your Cereal." Nearby, layers of decay and abandonment surround two lonely and stubborn houses.

We work our way past the hurricane-torn warehouses—one of them spray painted with the word "why"—and on to a half-filled parking lot. This area is filled with abandoned houses and buildings. Now it's a matter of finding the right one.

We zero in on two boarded-up and abandoned houses. We climb the first one's porch steps. A book lies open. So does the door. It's obvious that some-one else is currently squatting here. There are piles of belongings, including blankets, mattresses, pillows, clothes, and tools carefully arranged in one room. Another room is filled with printed material, some of which seems to be of a religious nature.

We decide to ride further, eventually passing an urban garden where Habi-tat for Humanity rents plots for a dollar per year. Some people use the garden to grow subsistence food; others grow produce to sell to local restaurants. A few local businesses own some of these plots. Another urban garden sits on the right as we ride beneath the North Galvez overpass, under which mat-tresses and other refuse litter the weeds. Graffiti on the overpass: "Turn the world around." Another building's graffiti reads, "DO NOT DEMOLISH: New Owner—William McGowan Notice Violation, 6/5/2015," in three separate places. Three sofas are stockpiled out back, sitting next to an enormous, semi-gated, and empty plot of land with a sign: "Private Property, No Trespassing." A train screeches nearby; we find more vacant lots and two more boarded-up houses on the corner of Montegut and North Miro streets. Piles of mattresses, old tires, and plywood litter the streets and empty lots. A lilac tree grows defi-antly above this mess. One sign lying in the dump: "Re-Elect Judge Medley."

Passing North Dorgenois Street, we spot a vacant house, perhaps suitable for unlawful squatting. Taking a left on the ironically named Law Street, which dead-ends at the train tracks, we discover that it is one of a cluster of four shambled houses worth investigating. Two are occupied. Of the other two, one is a gutted double shotgun—no furniture, no drywall, no anything, really, but exterior walls and, thankfully, a fine intact roof. The only thing living here are happy birds nesting atop the sad, drooping ceiling fans. Unfortu-nately, it sits next to a reasonably maintained, fully occupied house. The neighbors are too close, rendering our potential squat unsuitable. Shannon explains that good squats must be at least one or two lots away from occupied homes. It's too troublesome, and potentially dangerous, to squat next to neighbors who might call the police—or worse. Across the street and to the left, the remnants of another possible squat are hidden within runaway weeds and untamed shrubbery; this is a neighborhood nature is trying its best to reclaim.

Picking up our bikes, we realize the street doesn't dead-end. Rather, it curves right onto Law and Press streets, heading away from the train tracks, where a barbed-wire fence "protects" it from freight hoppers. Three stone steps emerge from the weeds on the right, with no obvious purpose. Perhaps before Hurricane Katrina devastated the neighborhood, there would have been a suitable squat here. If there was, it's long gone now. We hit Press Drive and Lausset Place and find a potentially perfect squat tucked away and hidden on the left, with the train tracks just ahead. On both sides sit vacant lots, the nearest house roughly thirty yards away. It's a crazy little green double-shotgun house, the property lined with tires and makeshift fences. A purple and green sign reads "NO DUMPING" in faded gold paint. We set our bikes near a rotting gazebo on the property. Vegetation completely consumes the front of the house along with its porch, blocking entry into the space. We find an alternative entrance—a hole in the wall on the right of the house, only accessed after traversing mounds of garbage, weeds, and roof tiles. We push through the weeds and climb through the hole to find solid floorboards and a suitable roof. In a pinch, the place is good for squatting. It is hidden on a barely navigable road that dead-ends at remote train tracks in the middle of nowhere. This is not the New Orleans I know or recognize.

In the distance, the skyline of that more familiar New Orleans peeks out from behind the tracks. Shannon says she knows of a better spot, so it's on our bikes, and we're off again. We take a left onto Press and hit Florida Street, where we find men working, surveying land, and operating construction machinery. The road is barely navigable, even for bikes, but the construction workers pay us no mind. The humidity is oppressive and feels punishing—the heat index kissing the 102-degree mark. In the distance, we see a building owned by the Sewerage and Water Board of New Orleans. It appears to be a rundown, dilapidated old pumping station, but we are unsure if it is still in use.

As we travel, we see scores of vacant lots peppering the half-abandoned neighborhood. Squatters sit on the porches of deserted homes, and feral roosters cross the proverbial road. It feels like a strange rural New Orleans just days after the storm, not ten years later.

We take a right on Comus Court to inspect a possible squat house. This decrepit pink house remains shut tight, with boards blocking any entrance from the windows or doors. A FEMA glyph marks the front of the house, a remnant from those dark days in the immediate aftermath of Katrina, signifying that the house had been searched by officials. Like those painted on nearly every house in every flooded neighborhood, the glyph is formed from a large "X" with a different notation in each quadrant. At the top, the number "926" signifies the date the house was searched: September 26, 2005. The lack of a mark in the right quadrant means no hazards were present. A "o" at the bottom means no bodies, dead or alive, were found inside. The "FL-2" on the left identifies the search party. We turn around and hit Rocheblave Street, heading down toward the train tracks to Port Street. There we see a variety of vacant houses, all offering possible squats. One house, gutted and half-rebuilt, has the single word "Tried" spray-painted on the front. The owners must have run out of money. Down half a block, we find two more boarded-up and vacant houses on the left and a warehouse to the right, all covered in graffiti. Next to the warehouse, we find piles of tires and wood near the shell of an old brick house—and graffiti that reads: "1 DEAD 9/17."

Ms. Lolita, an elderly woman who moved to New Orleans in 1976 and now lives next door to our potential squat, later told me that Pastor Joe William, a preacher from a church in the Ninth Ward, drowned in this house. According to Ms. Lolita, he was found "standing in the door stiff as a board" despite her pleading for him to vacate the premises ahead of the storm. It was a shame, she said. The preacher had seven children, or to capture her beautiful New Orleans pronunciation, "chirren."

We tramp through vegetation and over tires dumped in front of the house where the preacher once lived, and we make our way inside. We find a sign that reads, "Moms Drunk Again" and "Keep This Squat Nice Bitches." This "nice" squatter space has a sofa along with two chairs, in front of a fireplace. The room is littered with debris and used-up urban camping gear. Dozens of old, empty candles barely provide enough light to augment the bright light from the full moon. We walk through two rooms and, stepping through a hole in the wall, into the side yard of another house, boarded up except for a back opening where a door was once placed. This is where we enter our squat for the night. Shannon calls it "bum central." People she knew once squatted here.

At about seven in the evening, before settling in for the night, we search dumpsters and trashcans for cardboard-box bedding. We search other vacant houses looking for remnants of useful squatter gear. Graffiti on the back entrance of one squatter house reads, "Squat the World," but it has no cardboard boxes. We walk into another former squat; it's filled with roaches skittering and flying all over the place. We find enough flattened, and likely previously used, cardboard boxes on which to lay our heads for the night. We've brought along cheap whiskey to chase with our red wine. Together we return to our house to crack open the bottles.

Shannon shares a story about love and lust, dreams and fantasies, and about the joy of living a life of freedom—as well as the terror that such a life brings—as we drink the night away. It's a story similar to those of many squatters living transgressive, transient lifestyles deep in the underbelly of America. This world is not for the faint of heart.

GUTTER PUNKS IN NEW ORLEANS

It's common to walk past transients and gutter punks, or nomads, as they often prefer to be called, while passing through Decatur, Royal, St. Louis, and Frenchmen streets in and around the French Quarter. They might commit olfactory terrorism on passersby, but society treats them even worse. In 2015, the New Orleans Police Department targeted gutter punks throughout the French Quarter and Frenchmen Street, arresting twenty-two people on twenty-eight charges. In most cases, they were petty misdemeanors like aggressive solicitation, illegal vending, obstruction of public rights-of-way, carrying open glass containers, and public intoxication.[1] For these crimes, the names of each of those arrested appeared in the local paper—a typical Goffmanesque "degradation ceremony" that publicly shames and symbolically strips people of their roles as legitimate members of society and places them into the category of outcasts.[2] Every New Orleans police officer knows that these laws are rarely enforced, except as an excuse to question, arrest, or detain a person that garners the interest of a police officer.

The notorious local entrepreneur Sydney Torres, who made his wealth as the founder of a sanitation company called SDT Waste & Debris Services, a firm that scrubbed the city in the years after Hurricane Katrina, personally funded a private police patrol for the French Quarter and surrounding areas called the French Quarter Task Force. Residents and business owners can contact this private police force, which includes NOPD officers, using a cell-phone crime-reporting app.[3] Like his efforts after the storm, Torres's more recent labors are aimed at sanitizing the streets of the French Quarter and the Marigny. The difference is that, in this case, the garbage he's collecting is of the human variety. Most business owners in the area loved it. According to newspaper reports, city officials called it "a coordinated effort to address recent nuisance violation complaints from business owners" targeting transients, or gutter punks, in the city. "Basically," Torres said unapologetically to the *New York Times Magazine*, "I'm handling crime the same way I did trash."[4]

While some gutter punks can call home for money, many have burned those bridges after years of homelessness and drug abuse and now must confront the reality that they are not merely down but also out. These transient nomads lead resistant lifestyles and often pay the price for it. This distinct aspect of late capitalism—that it values humans based on their economic contributions to a capitalist labor market—is best revealed when looking at how these transients find the cracks and crevices of social space beyond the eyes of Big Brother and the system of social control.

THE SQUATTERS OF NEW ORLEANS

Squatting is the act of taking over an abandoned building and repurposing it for public or private use. Hobohemia, or the districts in the city where homeless people gravitate, consists of main "stems" or "drags" for down-and-out urban dwellers that are often geographically divided north, south, east, and west from city centers like the Loop in Chicago, Greenwich Village in New York, or the French Quarter in New Orleans.[5] Squatters in Hobohemia create "jungles," camps that act as social centers or places of leisure and rest free from hassle.[6] Urban campers, or squatters, make abandoned and vacant residential properties relatively secure spaces to lay their heads, safe havens away from tough urban life. These urban camps consist of largely improvised shacks near city centers. While some squats or urban camps serve as temporary habitats for the highly mobile transient populations moving through cities, many urban camps serve as what the classic sociologist Nels Anderson calls "continuous or permanent jungles" where squatters transform abandoned city spaces into safe places they call home.[7]

Of course, it's not always safe. Only a few years ago one of the deadliest fires in the city's recent memory took the lives of eight urban campers in an abandoned Ninth Ward warehouse at the corner of St. Ferdinand and North Prieur streets.[8] According to newspaper reports, most of those who perished were outsiders whose paths fatally intersected in New Orleans.

These young travelers are the latest version of Jack Kerouac's beatniks pursuing adventure, self-awareness, creativity, and transcendence. Some of these travelers make New Orleans their permanent homes and become part of the fabric of the local culture. Some even become locally successful artists and musicians or community advocates and activists. Others stay for a few days or weeks before heading out to new adventures on the road. Still others maintain annual seasonal residence in New Orleans after spending parts of the year hopping freight trains around the country or returning home to family. These travelers come in many varieties, from drugged-out gutter punks to educated, middle-class youths with an itch for adventure and travel, and everything in between. Some refer to these transgressive travelers as "crusty kids" when they are particularly dirty and overly tattooed, "travelers" when they are more "respectable" youths with some ambition, "hobos" when they get beyond youthful appearances, and "oogles" when they conceal their access to money. These drifters hop freight trains and hitchhike, making money by panhandling, begging, busking, and working odd jobs in the formal and semiformal economy. While some split rent in apartments, most squat in the abandoned houses all over the city, especially hurricane-torn houses in the historically black neighborhoods near the French Quarter. Urban campers turn these storm-damaged houses into temporary homes. This requires developing some serious urban camping skills, which many travelers learn while on the road.

TRAVELER NETWORKING

Many urban campers are travelers who develop creative networking strategies while vagabonding through the United States and Canada. They often connect in the most well-known bohemian districts and neighborhoods of the country—Bushwick in Brooklyn, the East Village and Lower East Side of Manhattan, Haight-Ashbury in San Francisco, Venice Beach in Los Angeles, and Wicker Park in Chicago, as well as the more "bohemian" areas of Seattle and Portland, Oregon—while also navigating to the new and upcoming hipster spots such as Savannah, Georgia, and Asheville, North Carolina, as well as various towns and cities in Colorado, a state where cannabis is legal. New Orleans is perhaps the new mecca of hipster bohemia, and it attracts transgressive travelers and gutter punks from all over the country. While some hitchhike across the North American landscape, many travel by hopping trains, using "hobo gold" as their guide, if they're fortunate enough to find it.

The travelers share with one another important information on travel and advice on the best places to venture. The best places usually share some common characteristics: a culture with a rich history; a large music and art scene; a bohemian reputation; relaxed or loosely enforced laws on loitering, busking, and panhandling; tourist destinations; cheap and affordable living conditions; opportunities for cheap and free food and drinks; and an established reputation for tolerance for travelers. Kia, a traveler from Washington and Vermont, learned about New Orleans and its music scene from gutter punks traveling through Burlington. She hitchhiked and rode freight trains to New Orleans. She claims her soul has a personal connection with the city. "If you are here, you can feel *it*." Still others have different but related stories about coming to New Orleans.

Gigi explains her reason for being in New Orleans using words such as "serendipity," "happenstance," and "chaos"—things that resonate well with conceptions of New Orleans. She grew up around "punk circus freaks" and now lives around the Mudlark Public Theater, a Port Street puppet playhouse. After hearing about New Orleans from fellow travelers, she arrived for "two magical weeks" following a particularly tough chapter in her life: the tragic loss of her boyfriend of three years, whom she describes as a "world-traveling freak clown," who died after falling out of a New York City building. They two had shared a deep appreciation of the English fantasy author Terry Pratchett's *The Colour of Magic*, which makes references to the colors purple and green[9]—colors that, with gold, make up the traditional colors of Mardi Gras and that, as she soon learned, turn up often in New Orleans. "Everything is purple and green, and purple and green flags are everywhere with clowns on it," Gigi said. "Just look at this magical place." She soon moved to the city of purple and green (and gold, too) and, as she puts it, "adopted the '*le bon temps rouler*' attitude of New Orleans."[10] Gigi explains, "We are just musicians trying to meet like-minded people," and advises, "Fail your ass off. Live your life and have real experiences that are much more important than pieces of paper that tell you how valuable you are." While sitting just outside of Fair Grinds Coffee near the corner of Esplanade and Ponce de Leon, Gigi proclaims enthusiastically, "I just want to live . . ."—she pauses as we both watch a man ride past on a ten-foot-tall unicycle—then she points to him and continues, "A life that is just like that guy."

Travelers like Kia and Gigi met while traveling the United States. Once they arrive at a stopping point, especially in cities like New Orleans, they use the same networking strategies to survive on the urban streets. One of the most important skills to acquire is finding a cheap place to sleep. Meeting up with urban tribes of squatters offers an opportunity to find quickly the most affordable temporary homes near the French Quarter and bohemian sections of the city—and thus a convenient entrée to tourist-industry jobs in the formal and informal economy.

URBAN CAMPING STRATEGIES

To be counted as a housing unit, the census defines that the residence must be protected from the elements, or covered by a roof, doors, and windows. This accounts for much of the increase in vacant housing from 2000 to 2010 because of the number of houses abandoned and boarded up after Hurricane Katrina. The 2000 census, taken five years before Hurricane Katrina, shows that there were 26,840 vacant houses and apartments in the city, or 12.5 percent of the total residences.[11] In 2010, five years after the storm, there were 47,738 vacancies, or 25 percent—an increase of almost 21,000 housing units. According to the Data Research Center,[12] the biggest changes in vacant housing occurred in Central City, which had 3,449 vacant residential units (a change of +1,252); Hollygrove, with 865 (+539); the Seventh Ward, with 2,641 (+1,385); St. Roch, with 1,574 (+710); and Little Woods in New Orleans East, with 3,583 (+2,942). In 2010, there were also 978 vacant residential units in the Lower Ninth Ward, 1,984 in the French Quarter, 735 in the Bywater, 1,052 in the Lower Garden District, 1,124 in Tremé/Lafitte, 1,821 in Mid-City, 554 in Uptown, 1,733 in the St. Claude area, and 478 in the Marigny. According to a 2014 Census estimate, there were 269,584 vacant houses citywide, or 13.6 percent of the total residences.[13] There were also 32,865 new houses built in 2010 or later, which could account for the lower percentage.

While census data provides information on the numbers and percentage of vacant houses in sections of the city, most travelers learn about the best places to squat from their social networks. In New Orleans, the best places to squat are in the historically black and/or working-class neighborhoods with a high percentage of vacant housing, like the Seventh and Ninth Wards and St. Roch.

Urban camping involves a hunt. After finding the target neighborhood, you must traverse these sometimes dangerous and unfamiliar areas, making sure to avoid the attention of neighbors. That's not always easy for white transients when traveling through predominantly black neighborhoods. The best way to travel is on a bike. It's often best to scope out numerous potential squats before deciding on the best available. The tools necessary for urban camping include a tent, cardboard box or mat, flashlight, lighter or matches, insect repellent, and tarp, but the most important tool is knowledge of the strategies involved in searching for a squat and identifying the best places to urban camp. One squatter explains:

> So another thing baffles me: Why don't people learn how to do this? We're not gonna be uncomfortable tonight. If it rains, the sun's not gonna beat down on us at 6 in the morning, police aren't gonna wake us up. It's like with the most minimal amount of forethought, urban camping anywhere is possible. . . . Just like in cities that are crawling with cops. You gotta find a spot where no one's gonna see you in the morning.

Space Between Neighbors

It's best to avoid squat houses that sit next to an occupied residential home. Neighbors will usually call the police if they suspect people are squatting in a vacant home next to their own. One reason is noise. Squatters don't want to have to remain in near silence while "at home." It's no fun to curtail behavior based on a fear of neighbors calling the police. And no one wants to get a surprise visit from the police in the middle of the night.

Intact Roof and Walls

The best houses for squatting protect urban campers from the outside elements, especially rain and the brutal Louisiana mosquitoes. Houses plagued with ants, mosquitoes, spiders, roaches, and other vermin can have severe repercussions on the health of squatters, as can overly damp conditions. I've seen many urban campers with horrible rashes and sores from insects, especially spiders. Many travelers suffer from what they call "rot foot," which one gets when traveling long distances in the rain with inadequate footwear. Suffice it to say, it's difficult and unhealthy to sleep in the rain or in wet conditions. A good squat must keep urban campers dry.

Level Floors Without Cracks

Squats with level floors make it easier for urban campers to set up a tarp and mat for the most comfortable night's sleep. Cracks in the floor give insects and other outside elements easy access to sleeping spaces in the house.

Back Yard

Don't shit where you eat, literally. Urban campers need a squat with a backyard to use as a bathroom. They usually piss in a specific area of the backyard and discretely crap into a plastic grocery-store bag and dispose of it in a nearby trashcan. One squatter put it:

> When I have to piss tonight, do you know what I'm going to do? I'm going to wrap my left hand around this beam and I'm gonna lean my ass off of here and then take a piss, and I'm gonna wipe with the toilet paper that I bought. And if for some godforsaken reason my body decides that I need to take a shit while I'm here, I'm going to use one of those plastic bags that I brought and the toilet paper that I bought, and that will be fine.

Escape Route

It's best to have a secondary exit in the event that an escape must be made from people with violent intentions, hostile fellow squatters, or police. Escape routes also allow squatters places to flee other dangers, including fires.

Inconspicuous Areas or Tolerant Neighbors

While some neighbors might tolerate squatters they deem respectful to the area, most will alert the police if they suspect people squatting in the buildings near their home. Since most squatters often travel in groups, packs of transients roaming in neighborhoods in which they obviously do not belong tend to attract the attention of community residents. Squats in inconspicuous places also make it easier to move in furniture and other household items that make urban camping more comfortable. As a result, the best squats are in inconspicuous locations removed from the eyes of residents or in places where neighbors tolerate transients roaming their neighborhoods. Though dangerous, some of the best squats lie in socially disorganized neighborhoods with high rates of crime, poverty, and other forms of social decay.

Close Proximity to Tourist Areas

Squats in close proximity to the French Quarter best meet the needs of many urban campers who make their living on the streets of the Vieux Carré. Many squatters lack reliable transportation and have a hard time keeping a bicycle safe, which makes cross-city travel difficult and sometimes dangerous. In fact, New Orleans has one of the highest bicycle death rates in the United States. It has become so bad that an organized group called the Bad News Bike Club paints white "ghost bikes" to place at each location in the city where a cyclist is killed, to memorialize the dead and serve as safety warnings for cyclists and motorists. Some of the city's miscreant youth "have fun" with what they perceive as hipster bike riders traveling in their neighborhood, sometimes threatening and attacking them as they ride by. Suffice it to say, living close to the French Quarter is more convenient and safe. Squats near the French Quarter, however, are becoming increasingly difficult to find. One squatter puts it:

> At the end of the day, there's only a certain amount of abandoned buildings in New Orleans that are in a close proximity to the French Quarter, where everyone needs to be to make money. So because of this, and because they're fixing more and more of these abandoned properties every year, and because more travelers come here every year, slowly, you're having an abandoned house population close to the Quarter that's dwindling, and

then you have a crusty population expanding. So these people want houses, except, crusty—the travelers, the punks, the people here who squat—whether they're clean-faced travelers, whether they're freshies, whether they've been riding trains for ten years, no matter who they are, there are still a limited amount of squats in close proximity to the Quarter.

Amenities

The more amenities a squat includes, the better the urban camping experience. While unusual, some urban camping sites might have running water, adequate plumbing, and a useable stove. Squats with private rooms and multipurpose rooms also add to the comforts of squats. While most squats look like garbage-filled dumps, other squatters put love and care into their squats, turning a potential dump into a fairly respectable place that can look almost like a regular, middle-class residential home.

PERILS OF SQUATTING

Make no mistake, squats can be dangerous. As a result, you've got to pick your monsters. There are unwelcome human intruders, like robbers and burglars, rapists and killers, police, and vigilante neighbors. Other monsters are of the nonhuman variety, including spiders, roaches, rats, mosquitoes, and other insects and rodents that crawl and bite. In many of the squats I frequented, roaches and spiders relentlessly swarmed the area while mosquitoes buzzed around like filthy, flying syringes. Many of the squatters reported—and provided visual evidence—of their rashes, sores, and infections. One squatter explains in depth:

> I'm well versed in bugs and spiders, like that's my thing. Oh, but with a spider bite, when it bites you it'll turn black in the center after it builds up and pusses, and then it's too late. You have to go to a hospital because that hole will keep continuing to expand. When you get bit by a spider, a day later if you've got a head on it, you take a needle or a knife and you pull that head off and then they have a white spot called the core. You've gotta take that core out because as long as it has a core in it, it'll continue to rot. And there are two, one for each fang.

While squatters view nonhuman dangers as a mere unpleasant inconvenience, most fear human predators.

Squatters tell nostalgic stories of a past when transients roamed North America freely with fellow drifters who helped one another out along the way. They provided advice on cities, train-route information, and dangers to avoid. This hobo intelligence was mainly communicated with glyphs scrawled on the fences of friendly homes and with graffiti indicating the conditions and safety of the squat. They also followed informal rules, a sort of hobo code of conduct that travelers, it is widely believed, recognized and respected. In fact, a formal hobo code actually does exist.[14] Hobos even have their own annual meeting, the Hobo National Convention, which was established over a century ago.[15] At the 1889 National Hobo Convention held in St. Louis, a strict ethical code was established for hobos to follow while on the road. The first of the sixteen codes, "Decide your own life, don't let another person run or rule you," is a hobo philosophy that many transients still share. As one traveling youth in New Orleans put it: "Freedom means I do what I want to do when I want to do it." The other codes set the ethical guidelines for the transient community. A few of these codes follow:

- When in town, always respect the local law and officials, and try to be a gentleman at all times.
- Don't take advantage of someone who is in a vulnerable situation, locals or other hobos.
- Always try to find work, even if temporary, and always seek out jobs nobody wants. . . .
- When no employment is available, make your own work by using your added talents at crafts.
- Do not allow yourself to become a stupid drunk and set a bad example for locals' treatment of other hobos.
- Always respect nature; do not leave garbage where you are jungling.
- If in a community jungle, always pitch in and help.
- Help your fellow hobos whenever and wherever needed, you may need their help someday.[16]

The sociologist of "Hobohemia" Nels Anderson found that in the 1920s, while urban campsites—or "jungles" as he called them—were rather hospitable and egalitarian, they had an informal code of conduct, or etiquette, that had to be adhered to strictly in the face of potentially harsh consequences. This unwritten but strictly enforced code of conduct addressed "jungle crimes," including making fire by night in jungles, robbing other squatters at night, making the jungle a permanent hangout for "buzzards or moochers," wasting food, and leaving pots and utensils dirty.[17]

These rules now seem to many of the transient youths a more romantic and nostalgic relic of days long past. Now, they argue, the transient life of squatting in urban centers is fraught with danger. Most transients account for this change in behaviors as both a "cultural shift" related to the increased demands of squat houses and the perceived increase in drug use among the traveling population. One urban camper explains: "There's less places to stay that are close to the Quarter and there's more scumbags that are staying in these places. So good people are forced to do bad things to make sure that their pack isn't getting stolen, and it's not getting any better."

Other squatters echo the same arguments, stating that more transient youths are flooding into the city, and that while most respect the informal rules, others act irresponsibly mainly because of to drug use. One squatter says, "More kids are coming every year, and half of these people are like me: They have a hustle on the street, they make their money, they eat their food, they swipe their cigarettes. . . . And then the other half do heroin. And those kinds of people can't coexist in these places [squats]." Some squatters complain that transients looking for an urban camping squat will come across their gear and steal it, a clear violation of formal and informal hobo codes.

Another squatter puts it, "And then people run off with each other's bags. And then people get into verbal arguments, and there's just animosity now. The cultural shift is also happening in New Orleans because people have to be cutthroat, but they don't want to be, necessarily. But they're being forced to." Still another transient youth squatting in the city argues that the desperation of the poor is reaching new heights. She says, "The amount of people that are aggressively panhandling in the city? They beg me. They beg *me*! I'm standing on the fuckin' corner, sitting on my crate! But you know what I'm saying?"

Others refrain from providing explanations of why things appear to be more dangerous. Instead, they describe the actual dangers. Louis, who has been squatting for the past few months in New Orleans paints a vivid picture of one of his most frightening nights in an abandoned building on North Rampart Street between the French Quarter and Faubourg Tremé:

One of the nights, I wake up to a bunch of flashlights. I'm used to sleeping in the road with my dog, like we don't like waking up to a bunch of flashlights. . . . Not only are there a bunch of flashlights, but they're swarming in through the front room and I start like, you know, looking around. And this little lazy dog. . . . I look around and like, all of them are running around with needles. Like, strapping, getting their veins ready, like literally running around with needles. In my head, I'm just like, I froze. Cause like, if one of these fucking people fall, that could be a needle in me. That is not something I was okay with. Like, having to wake up with that? Like knowing that my dog was running around on the floor, and they were like running around stumbling everywhere? There's a level of comfort and respect, and they danced all over that.

Another transient youth who has been traveling for the past two years explains how things sometimes go down among squatters:

Except nowadays, when that happens with people, when people have a spit [argument], when somebody fucks up, and somebody decides they have to go and settle it, it's not about one-on-one anymore. Now, people come to where you live, your squat, with locks on chains, they call 'em smileys, and they'll beat you all upside the head with them. They'll come in when they know there are four of you in there. And they'll just beat you all up. It doesn't matter who's there. Stomp crew. They don't take your life. They want you to have your life because they want you to feel the pain. These kids are fucked up. *These kids are fucked up.*

One squatter sums it up nicely, without trying to explain the causes for the perceived growing dangers: "It's becoming scarier. I really wish I could tell you why everyone is acting so crazy. I can only offer those as my insights. It scares me." While many transients view themselves living a life of freedom, doing what they want and bowing to no master, it comes at a price. These vagabond wanderers might find creative ways to subvert institutional authority and avoid the normative expectations of mainstream society, but they also lead lives that often result in either self-destruction or tragedy at the hands of outside elements—both nonhuman and human. In the following case of the black political squatters of New Orleans, their greatest enemy is city government and its criminal justice system.

THE SQUATTERS OF WASHITAH MU'UR NATION

Some squatters of New Orleans don't just want to make a home. They want to make change. So they take over vacant houses to create urban camps that serve as their base for practical, spiritual, cultural, and political pursuits.[18]

Many black political squatters are part of the Washitah Mu'ur Nation, which combines many worldviews.[19] They follow the teachings of Noble Drew Ali, the founder of the Moorish Science Temple of America, who reimagines black identity and racial pride through developing alternative education and spiritual-reawakening methods that recast black history in a new light. He argues that black people are traditionally believers of Islam and are descended from the Moors. Along with following the ideas of Ali, they practice yoga at their squat house they call the Smai Tawi Temple.

Rota, a strikingly beautiful young woman of only eighteen years of age and one of the city's leading black political squatters, runs the Smai Tawi Temple. She claims that she lives the ancient Egyptian pursuit of life, liberty, and happiness. She says, "I am not a squatter but a national in America."

She explains further while showing me around her temple squat: "That's what we do, I was just doing yoga right here, laying on the mat doing Smai Tawi, getting my mind right and uploading the messages from the sun. Ascending with the earth." Smai Tawi, she claims, is the first word for yoga and means merging one's higher self with one's lower self. "So that's the purpose of this space, to bring synergy and mastery of ourselves and learning how to not be unbalanced. . . . We meditate at 4 a.m. when we wake up and say the forty-two laws of Ma'at Under Kemet Law [an Ancient Egypt source for balance and moral and spiritual instruction] and meditate."

The Washitah Mu'ur Nation is a group that claims to be descended from indigenous people of the Americas that were black Africans, or at least a sovereign tribe descended from pre-Columbian black people who settled in North America. The Southern Poverty Law Center likens the Washitaw Nation, as they spell it, to a sovereign-citizens movement of "free peoples" not subject to laws imposed by city, state, or federal governments. Washitah Mu'ur Nation of the Smai Tawi Temple also claims to have a right to the land. In particular, this claim applies to New Orleans, where they argue that the Louisiana Purchase of 1803 involved Emperor Napoleon Bonaparte selling only the city streets of New Orleans and its military barracks, not the approximately 827,000 square miles of land west of the Mississippi River, for $15 million to the American government.[20] The rest of Louisiana, they believe, was stolen from the Washitaw Nation. As we sit in the Smai Tawi Temple, Rota explains their political and social position:

> It's Smai Tawi. It is coming into consciousness that you were born on a planet that you don't have to pay fiat notes to live on. You, you came to your mama, did you ask "Mama, born me?" No. So why the fuck you gotta pay rent? That doesn't make no sense. That's not, that's not even sane. That's not sane. And for us to continue to live in an insane manner is fucked up. These people are gonna continue to kill each other if they keep living this way.

When asked what kind of movement they would consider themselves, three members of the Smai Tawi Temple squat agreed that it's a divine and national movement that involves spiritual, political, social, and economic dimensions. Rota says:

> It's all of it, all of it. Like I told you, when Prophet Noble Drew had hit the pinnacle of the solution, he was sitting at the round table with the Congress and the Senates and the states and all of this to claim back our vast estate, and then the fucking Stock Market crashed and they started playing the game to get people to sign birth certificates so they could get them as collateral. That's when this game started.

They also claim that their movement is matriarchal, with strong ties to what some might consider radical feminism—and just a touch of eschatology. Rota explains that women who exist in this patriarchal society suffer from "toxicity." As a result, part of their belief system involves "lifting up fallen humanity, lifting up the woman." Rota says, "Women are born in toxicity. We are giving birth to toxicity, a toxic mind state. Until women get their womb clean and figure out what is the natural order of this, it's not gonna be right." According to Rota, getting the womb clean involves mental, physical, and spiritual healing. "We all have spiritual damage. The system we live in is damaging us spiritually to where we are losing our souls. We're in the end times." While the power of matriarchy remains important, the black political squatters want to transform their squat temple into a youth-empowerment project.

Dongi, another one of the Smai Tawi Temple squatters says, "We're just working to make this a transitional living space for anyone that wants to learn how to live with rain water and solar power, and bringing the catchment system over here. We're actually getting certified as a 501-D communal nonprofit organization. This is now Smai Tawi. It's like a reservation."

The Washitah Mu'ur Nation of the Smai Tawi Temple also uses transgressive tactics to stake legal claims to abandoned or vacant houses in the Bywater neighborhood, deploying invented documents to present to police in order to avoid forcible removal from their squat, criminal charges, or both. In one of their squats, they issued to police a ten-page document, filled with seemingly nonsensical legal jargon, that laid "legal" claims to their rights to the property (see figure 6.1).

According to the *Times-Picayune*, one legal expert admits Washitah Mu'ur Nation's cleverness. One real-estate lawyer is quoted as saying, "In a way, the Washitahs' document is clever because it includes a legal description of the

THE MU'UR NATIONAL REPUBLIC
MU'UR DIVINE AND NATIONAL MOVEMENT OF THE WORLD
Mu'ur Americans, Aboriginal and Indigenous Natural Peoples of Northwest
Amexem, Northwest Africa / North America / 'The North Gate'

New Orleans, Louisiana Republic   January 13, 2016

Yahmel Yaffu Ali Bey   Department of Safety and Permits
AND   Historic District Landmarks Commission
Yanamaria Latasha Bey   Vieux Carre Commission for Historic, Zoning,
                              And Building Violations
VS   1300 Perdido St., Room 7W03
                              NEW ORLEANS, LA REPUBLIC [70112]
CITY OF NEW ORLEANS

NOTICE OF CHARTERED EJECTMENT

Notice to Agent is Notice to Principal. Notice to Principal is Notice to Agent.

IGNORANTIAL LLEGIS, NEMINEM EXCUSAT
Ignorance of the Law is no excuse.
To Whom It May Concern:

The property situated at the addresses commonly referred to as: 3404 North Rampart Street, New Orleans, Louisiana Republic [70117], with a legal description commonly referred to as follows: Municipal District 3; Square 286; Lot 21 N. Rampart, 19 x 102, Latitude, Longitude (DMS) 29° 57' 55.1124" N, 90° 2' 26.5236" W, has been SEISED IN DEMESNEAS OF FEE, by Yanamaria Latasha Bey, who is part and parcel with the land, 'Authorized Representative of the Washitaw Nation', in an estate in fee-simple as a corporeal hereditament and now deemed the true lawful possessor of freehold property. Yanamaria Latasha Bey has seised as the true possessor of the land itself, with an estate of inheritance in fee-simple absolute. In the act of ALLODIUM peaceable possession, Yanamaria Latasha Bey is the true possessor bonafied finding the property in haereditas jacens prior to possession. Exercising rights in full capacity as in full life denoted as a Natural Person and an indigenous American National, Yanamaria Latasha Bey, lineal heir performed peaceable possession in the law of/La Ley favour 1/inheritance d/un home./

Bywater house, citations to legal cases and federal statutes, mysterious Latin words, and a whole lot of misused legal jargon. But when you look at what it actually says, it's completely incoherent."[21] But it worked, at least at first.

Despite the "legitimate" property owner—who is from California—showing the deed, bank documents, and tax records of the house to the police, authorities refused to evict the black political squatters, who seemed to have a legal document staking a claim to the property.[22] The *Times-Picayune* reports: "In explaining why the squatters were not immediately turned out, the Police Department said that it wasn't reasonable to expect a beat officer to parse seemingly official documents on the fly. The department, as Superintendent Michael Harrison put it, 'had to do its due diligence.'"[23] According to the same newspaper report, Louisiana does have laws providing squatters' rights. Though the article does not elaborate much on these rights, it is referring to Louisiana Civil Code Article 3486 of Acquisitive Prescription. In Louisiana, while criminal law is based on English common law, contract and tort law between private-sector parties is based on French and Spanish laws that derive from Roman law. Like so much else, even our legal system is different than the rest of the country. Louisiana's acquisitive prescription is a civil law analogous to adverse possession, sometimes called squatters' rights.

Of course, to possess a property that you don't own is no small task. Adverse possession requires a squatter, or adverse possessor, to treat a property as an owner would. The adverse possessor must accomplish the following:

> Act like an owner in her actual, open and notorious, continuous, and hostile possession because her possession should be sufficient to assure that the true owner has been provided with sufficient notice of the acts of the adverse possessor. . . . Actual possession requires that the possessor physically possess the property and have an intent to maintain control of that land. The possessor must exercise his dominion and control over the property such that there are visible signs of the possessor's occupation. . . . Open and notorious possession refers to possession that is apparent and visible, as opposed to possession that is hidden. . . . Continuous possession requires that the adverse possessor exercise acts of possession over the property throughout the entire requisite time period. . . . Hostile possession occurs when the possessor occupies the land without the consent of

the true owner. When the true owner grants the adverse possessor permission to possess the land at issue, possession is not hostile.[24]

Acquisitive prescription is encoded into the following articles of the Louisiana Civil Code:[25]

Article. 3424. Acquisition of possession: To acquire possession, one must intend to possess as owner and must take corporeal possession of the thing.

Article. 3446. Acquisitive prescription: Acquisitive prescription is a mode of acquiring ownership or other real rights by possession for a period of time.

Article. 3476. Attributes of possession: The possessor must have corporeal possession, or civil possession preceded [sic] by corporeal possession, to acquire a thing by prescription. The possession must be continuous, uninterrupted, peaceable, public, and unequivocal.

Article. 3486. Immovables; prescription of thirty years: Ownership and other real rights in immovables may be acquired by the prescription of thirty years without the need of just title or possession in good faith.

Article 3486 serves as the real kicker, requiring the adverse possessor to occupy the property in a "continuous, uninterrupted, peaceable, public, and unequivocal" manner for thirty years. Washitah Mu'ur Nation of the Smai Tawi Temple fell well short of that time frame, occupying three houses for only a number of months. They made a good run, but eventually the police arrested four members of the group at their squat on North Rampart in the Bywater. They pleaded not guilty to charges of criminal trespassing. Three of the squatters refused to sign a peace bond required for release and provided the names Batman and Sub Zero in municipal court.[26] The property owner, Fred Hines of California, according to the *Times-Picayune*, wants these young homeless people between the ages of eighteen and twenty-five to be prosecuted to the "full extent of the law possible." It looks like he will get his wish. Besides the charges of criminal trespassing, simple burglary, and the punitive charge of resisting arrest (for refusing to give their names or stating legally false names), at least one of the four now face felony burglary charges and up to twelve years in prison. Although it was a simple homeowner that called the police, arresting harmless young squatters and charging them with felonies that could imprison them for up to twelve years shows how the powerful use the law as a potent weapon to dispose of socially unproductive members of capitalist society or people that pose threats, however insignificant, to the established order.

This behavior is reminiscent of what the sociologist Steven Spitzer calls "social junk" and "social dynamite." Social junk, from the point of view of the dominant class, is a costly yet relatively harmless burden to society. The lack of creditability of social junk resides in the failure, inability, or refusal of this group to participate in the roles supportive of capitalist society. Social junk is most likely to come to official attention when informal resources have been exhausted or when the magnitude of the problem becomes significant enough to create a basis for "public concern."[27] Social dynamite actively questions "established relationships, especially relations of production and domination. . . . Social dynamite tends to be more youthful, alienated and politically volatile than social junk. The control of social dynamite is usually premised on an assumption that the problem is acute in nature, requiring a rapid and focused expenditure of control resources."[28] The Washitah Mu'ur Nation of the Smai Tawi Temple, composed of homeless idealistic youths, can fall into either of these categories, depending on one's point of view. Either way, the felony charges are a clear example of the state's attempt to remove transgressive social actors, or "problem populations" and unproductive members of capitalism, from society and lock them in cages. In this exclusive society, transgressive social actors are nuisances that cause problems for the powerful. The state serves as a vital weapon of the powerful to exclude these transgressive populations from the social world and its institutions.

According to follow-up newspaper reports, the member of the Washitah Mu'ur Nation facing felony charges, who calls himself Atum-Bey, grew up as a middle-class kid in my home neighborhood, Gentilly. He was, according to his mother, a gifted child who "won quiz-bowl-style competitions, played football, and displayed a kind of magnetic personality and independence that led other kids to gravitate toward him. His grades and ACT scores won him a full scholarship to Loyola University." He became intellectually curious about the world and began to "search for meaning to life's unanswerable questions." This is what led him to explore the many ideologies of various groups and cults as well as classic and New Age philosophies. Even the local newspaper admits that "authorities may also be trying to make an example of the Washitaw" and the practice of squatting, as "staffers in Orleans Parish say they increasingly have had to deal with an avalanche of sham filings" from homeless people trying to claim vacant or unused homes.[29]

Perhaps the social junk is getting pissed and transforming into social dynamite. As the powerful in our society continue to meet resistance from those who most acutely experience the structural problems posed by late modernity, and as people become increasingly disenchanted with a society steeped in the contradictions between its promises of freedom and wealth and exclusive institutions that grant such privileges to only a select privileged few, we can expect governments to play a larger role in dishing out harsh discipline to transgressive populations. Such is the case with the homeless youths of the Washitah Mu'ur Nation, who, despite their nonviolent claims to a home not of their legal possession, and despite the fact that nothing was damaged or taken from the home, face punishments similar to those handed out to the most violent members of our society. Atum-Bey's saddened mother understands the contradictions of the system's version of "justice," stating: "If the goal of the legal system is to correct bad behavior and redirect criminals toward productive lives, it is doing the opposite in this case. He's going to have to get a job sometime. Having a felony on his record is not going to help with that."[30] As the graffiti of a house in the Mid-City neighborhood says: "Homeless people plus empty buildings equals society fails."

A NIGHT AT THE HOMELESS SHELTER

Given that homeless shelters exist, I wondered why some people choose instead to sleep underneath interstates and in filthy, dilapidated squats.[31] The excerpt below stems from my experiences spending the night in a homeless shelter near the Central Business District in New Orleans.

Near the corners of Clio Street and Oretha Castle Haley Boulevard, a sign points to the check-in building for the New Orleans Mission, a local homeless shelter. At around 4:30 p.m., people begin showing up in numbers. Only about two blocks away, beneath the elevated Pontchartrain Expressway, dozens of people prepare for another homeless evening with scores of their compatriots. Some down-and-out folks arrive at the shelter for an evening meal and return to the nearby underpass; others arrive for a night's rest away from the outside elements. A line of men queue for admittance into the shelter. A guy with missing teeth pats down the bedraggled men like a cop during an arrest.

While looking at a sheet of paper containing some sort of list, he starts asking questions. First he requests my ID, which I don't have. "What's your name?" he asks. I tell him, and for some reason he begins checking a four- or five-page packet with names on it. He tells me to lift my arms and begins to pat me down, touching my chest, stomach, legs, ankles, and feeling around the pockets of my jeans. Satisfied, he points the way forward. I enter the homeless shelter through a storage area, down past the women's bathroom and a kitchen, and head into a large cafeteria area—a converted warehouse, really—with long, white folding tables and black chairs.

A man sits behind a desk with a computer where he processes people for a bed ticket. He asks for my name, Social Security number, birth date, and if I have taken a tuberculosis test. He informs me that I must get the test done within a week in order to remain at the shelter. The black man behind the desk enters my information into a computer and writes out a ticket. He hands me a questionnaire and invites me to sit at a table on the side of a cafeteria to wait for an entry interview.

I enter an office where three people sit behind computers, asking people questions and writing answers on a form. Those forms are later entered into a computer database. The man asks me for my name, my reason for needing a homeless shelter (as though that wasn't already obvious), if I am on parole, was recently in jail, am former military, or am addicted to drugs and alcohol. I provide my name and reply "No, no, and no" to the other questions.

Then we arrive at the religious part of the interview. He asks if I believe in God. I said I don't know. He asks if I died today and the Lord appeared before me and asks me why I should receive entry into heaven, what would I say. "I'm not sure." He asks, "So you don't believe in God?" He instructs me to wait outside at the white folding tables while he enters the data into a computer. After about five minutes, they call me back into the office. Another man tells me to sit at his desk and begins to ask more questions about my faith in Jesus. He drills me about God, why I don't believe, the importance of eternity, and so on. This was all an attempt to get me into a twenty-one-day program. It reminded me of Orwell's observation: "It is curious how people take it for granted that they have a right to preach at you and pray over you as soon as your income falls below a certain level."

After about fifteen or twenty minutes of questions and a lecture on the importance of faith, he tells me to sit next to another man, who asks even

more of the same questions before taking my picture. He tells me that everyone must take a shower before going to bed and asks if I need clothes. He
hands me a packet that contains cotton swabs, a razor, face towel, the New
Testament, lotion, a toothbrush, toothpaste, soap, and finally, a bed ticket. He
directs me into a makeshift chapel, where a full-blown religious scene unfolds,
reminiscent of one that would take place in a Charismatic storefront church
in the inner city.

Thumping, religiously inspired Christian music fills the room, now packed
with both men and women dancing and clapping in jubilation—some people
even get the Holy Ghost and speak in tongues. A huge black man plays the
keyboards while a skinny white guy plays guitar. We all sit in this church service, listening to the pounding music in anticipation of the preacher about to
deliver a sermon. Unlike Orwell's experience at Christian-influenced homeless shelters, many people heed the church service with unadulterated enthusiasm. They shout "amen," "yes," and other cries of affirmation. Others sit tired
in their place, calmly and solemnly waiting for the service to end. The music
lasts for over a half an hour.

The chapel space consists of a makeshift room in a warehouse with sets of
pews facing an altar composed of a podium with space for musicians and a
lectern. One sign in the chapel says, "Please sit up, no laying [sic] down." A
preacher arrives at the scene with an uplifting sermon about how all the
people here will become future leaders, teachers, service people, and preach-
ers who have yet to realize their potentials because of a stranglehold placed
upon their lives. After an altar call during which three people approach
either for prayer or salvation, more music plays. The service finally ends at
about 7:10 p.m. The women exit the chapel on one side, men on the other.
The church service is the only activity in the homeless shelter both sexes
share. While the women leave to eat at another building, the men make their
way to the cafeteria. We wait in the dinner line for about ten minutes. The
men stand sadly, their faces long and their stomachs empty. They say nothing to one another, looking either at the floor or into the distance.

Most of the men leave immediately after dinner. About fifty men remain
to sleep in the shelter. At about 7:45, I walk around to the outside of the facility
near the weights area. One of the workers scolds me, saying overnight guests
are not allowed to walk around the facilities. He orders me to sit in the chapel
area with the rest of the men.

Most of the men blankly stare at a television screen, sitting in the pews in the same space that was just a few minutes ago a church service. At 8 p.m. a man yells, "Smoke break!" I walk down the steps with the men heading outside. The smoke break lasts for about ten minutes. It's the only time the men are allowed to congregate outside or have a smoke before bed. After smoking, the men return to the chapel room to sit in the pews and watch the Nicholas Cage movie *Bad Lieutenant: Port of Call New Orleans*.[32] Most sit silently, almost evenly distributed across the chapel area. Some read, but most watch TV. Some talk, but most are silent.

In order to sleep at the shelter, one must present a bed ticket to a night-time bed monitor with the word "approved" highlighted in yellow on the back of the ticket, next to the day's date. It is only possible to get this word "approved" highlighted (aside from smuggling in your own highlighter) from a shower monitor, who admits the men into the showers and provides them with a towel. The shower monitor hands me a towel and watches me enter the shower and bathroom area.

The community showers resemble the showers one would find in prison. About six naked men stand together in close proximity underneath shower-heads, most wearing sandals or flip-flops. I have no sandals or anything to cover my feet and feel a slight revulsion about stepping with my bare feet in the showers. Instead, I spend about twenty-five minutes washing my face, wetting my hair, brushing my teeth, and other ablutions until enough men have come and gone that no one will notice that I avoided the shower. Before leaving the shower area, I make sure to drench my face, hair, and chest in water while walking out rubbing a towel over my head as if I just finished bathing. The man highlights "approved" on the back of my bed ticket.

Showers shut down at about 8:30 p.m., bedtime starts at 9 p.m., and lights out at 10 p.m. There are about seventy-five single-sized bunk beds on the second floor of the warehouse. First, a man with a list of names asks for your bed ticket. He looks over his sheet to find your name and makes sure the back of the ticket says "approved." He asks me what is in my plastic bag. I reply, "What y'all gave me." Most of the men lie in bed sleeping; some stare at the black ceiling and the dim fluorescent lights. When you get to your bunk you must put on the mattress cover and pillowcase. Some men sit up, many coughing; others toss and turn. One man reads on his laptop. One man says he did twenty-two years in the Louisiana State Penitentiary at Angola and that homeless shelters offer about the same experience. At about 9:45 a man with a clipboard looks around to make sure we are in the right beds. The lights shut off promptly at 10 p.m., leaving the room nearly completely dark. We are left with our private thoughts while lying awake and listening to the sounds that dozens of men make throughout the night. People begin to wake up at 5:30 in the morning. Breakfast begins at 6 a.m., and an hour later everyone must leave.

Sleeping at homeless shelters is highly structured. Once a homeless man enters the shelter—which must be by 6 p.m.—the workers completely regulate his life. One feels worthless in a homeless shelter and, worse, beholden to others who can "save" you. There is nothing more revolting than getting help from people who fancy themselves your saviors. In many ways, we are fortunate that shelters such as these provide homeless men and women a place to eat and sleep. But I'd prefer a night beneath the overpass or in a squat any day.

HOMELESS PEOPLE + EMPTY BUILDINGS = SOCIETY FAILS

Scores of homeless people, transient youths, bums, squatters, skells, vagabonds, and wanderers exist in the urban fringes of postindustrial New Orleans. I encountered dozens of homeless travelers, from Roma drifters to train hoppers to old and tired black bums to drugged-out middle-aged white folks to young homeless families with small children. I interviewed dozens of homeless youths in their twenties who sit all day in the grassy areas of Jackson Square or on its benches facing St. Louis Cathedral. There is even a woman who calls herself the "mother of Jackson Square" who acts as a semi-maternal figure to the many homeless street kids who roam the area. While some of these kids may suffer from mental-health issues, most seem to have mental faculties similar to most "normal" folks, even if they have a disdain for authority and rules.

Amid the tourist and entertainment zones of the French Quarter, swarms of homeless people exist in the background but in plain sight of the relatively wealthy tourists strolling Crescent City streets. Many of their sad stories involve destroyed families and shattered dreams, along with wishes, hopes, and unrealized aspirations. In these streets, you will find the stories of people like Suna and Liz, two homeless transients who drifted from one dilapidated house to another trying to find some normalcy in a city that denied them a warm welcome. Their squats were filled with garbage, filth, and crawling insects and pests. Barely in their twenties, they began to look, over a short period of time, as broken down and dilapidated as their squats. But they somehow maintained a stubborn resiliency, recording their wild and sad journeys on cheap cellphones—and they even fell in love and requested a gutter-punk wedding right on the Moonwalk overlooking the majestic Mississippi River. Liz says she is now pregnant. They have a friend named Louis, a local kid they left behind when moving out to, hopefully, better pastures. Louis is homeless but still hopes to attend the University of New Orleans one day. He now has a kid on the way, too.

Many of the young homeless lads turn tricks for homosexual men to earn money, and their street girlfriends take odd jobs, like learning to mime with people like Gold Man. Underneath the St. Claude overpass and other interstate overpasses in the city you will find tent camps where dozens of homeless people create makeshift homes and pockets of protection.

Meanwhile, thousands of vacant houses dot the streets of New Orleans. Many of these houses were abandoned following Hurricane Katrina and remain abandoned. No one knows what to do with these unclaimed properties, but most people intuitively understand that a city with an abundance of both empty houses and homeless people amounts to a failed society. Despite living in a so-called progressive era, a hatred for the poor remains prevalent. As Orwell put it, "It is fatal to look hungry. It makes people want to kick you." We hold on to the notion that poor people somehow deserve their fate. In 2014, Louisiana voters rejected a proposed amendment to the state constitution that "would have allowed for the sale of abandoned property following Hurricane Katrina for $100 per lot. Voters disagreed with the Legislature's plan to essentially donate the abandoned properties. The effect of the no vote, however, was that the abandoned properties remained abandoned."[33] The idea of giving free handouts to poor people is, apparently, more revolting than the squats and urban campsites of the homeless or the sight of rotting houses that once served as homes.

Considering homelessness a personal issue focuses attention on the individual lives of the many varieties of homeless types throughout the city, each with a unique and fascinating story yet to be written. There is no single event that typically leads to becoming homeless. Rather, a series of steps occur in a longer process. For most of us, we are only a few steps away from the same sad stories the homeless tell us about their lives. And when we tally up our own political and economic capital, most of us are socially and economically much closer to homeless bums than to the politicians for whom we vote. Yet we identify more with our politicians than our homeless brothers and sisters. The personal troubles of homeless people remain just that: personal troubles—even though that same fate may befall any of us at any point in our lives. But personal troubles do little to explain the structural issues that surround homelessness in the many makeshift squats all over the city.

When we find so many vacant houses and homeless people and transients living on the urban fringes in dilapidated squats, the problem goes beyond the personal troubles of milieu.[34] Rather, this is a structural problem where the disease lies in the very nature of our societal institutions. Ameliorating the disease requires operating at the institutional level, changing the composition of our society. It requires a cultural shift in our way of thinking, especially with regard to the value of human life. A society that threatens to build walls between countries, deports vulnerable humans without "papers," allows police officers to manhandle children and shoot unarmed black men, drops bombs on villages in the Middle East, claims that foreign policies that lead to the death of over 500,000 children is good for the country,[35] creates drug laws to destroy and undermine progressive thinkers and marginalized black communities, and, on top of it all, allows the social problem of homelessness to exist when houses remain vacant is, in short, a fucked-up world in need of massive and revolutionary change.

CHAPTER 7

OCCULTISTS AND SATANISTS

The Paris slums are a gathering-place for eccentric people—people who have fallen into solitary, half-mad grooves of life and given up trying to be normal or decent.

—George Orwell, *Down and Out in Paris and London*

ONE WAY OR ANOTHER, magic is real in New Orleans. Part of the city's cultural life involves the nocturnal outpourings of Voodoo, occult practices, and Satanism, as well as the scores of other religious practices that take place in our cemeteries and other religious sites. Beyond the spiritual tourism of legendary Voodoo Queen Marie Laveau's burial site, specialty Voodoo and occult shops, the New Orleans Historic Voodoo Museum, Priestess Miriam Chamani's Voodoo Spiritual Temple (just below the apartment where Zackery Bowen dismembered the body of his girlfriend Addie Hall in October 2006 and placed her body parts in the refrigerator, oven, and kitchen pots),[1] haunted New Orleans tours, and all that Spiritualist yaya lies an underground New Orleans that includes a subterranean world where occultism, Voodoo, and witchcraft quietly creep and crawl in the swampy foggy nights. We also have gruesome stories that blend fact and fiction about colorful but disturbing and sordid characters, like the "Axeman of New Orleans"—a man who butchered at least twelve of the city's citizens and threatened to kill more if they did not play jazz music fifteen minutes past midnight on March 19—and the New Orleans Creole socialite and serial killer of slaves Delphine LaLaurie, or Madame LaLaurie, who chained, tortured, maimed, and killed her slaves in the basement of her mansion at 1140 Royal Street in the French Quarter. There exists the aboveground specialty shops such as F & F Botanica Spiritual Supply in Tremé, Esoterica, Mystic Tea Leaves, the Museum of Death, Marie Laveau House of Voodoo, Voodoo Authentica, the Island of Salvation Botanica, shrines of La Santisima Muerte, the Thelemic Ordo Templi Orientis Temple in Bywater, and of course Congo Square, where slaves once performed dancing and drumming rituals. More importantly, it's the private spaces and hidden areas in the city where much of the magic and many of the rituals happen in New Orleans—like the St. Roch Cemetery; the Barataria Preserve, outside the West Bank city of Marrero in Louisiana's wetlands; and in the private homes of witches, warlocks, occultists, and Voodoo practitioners. These city dwellers transcend the ordinary in, at least to outsiders, highly unusual ways. While they may not all be down and out, all of the practitioners of the supernatural in New Orleans occupy an important part of the cultural life in the city's urban underbelly.

Underground religion, spirituality, occultism, and supernatural belief down here in New Orleans cannot be explained as Marx's sigh of the oppressed or opiate of the people or as Durkheim's struggle to keep anomie at bay or strengthen the collective consciousness. Structural functionalists understand society as being composed of interdependent parts operating for the overall functioning of the system. They tend to view religion as having an important societal function that maintains group solidarity, integrates people into a normative order, or at least keeps people in their place while they hope for a better existence after death. The conflict theorist, on the other hand, sees all the parts of the system, or the institutions of society, as vehicles by which the elites impose and maintain their power over the masses. Those who gravitate toward the conflict perspective would likely argue that beliefs in the supernatural maintain the ideology of the powerful group's position over the people. Some scholars might claim that supernatural beliefs, such as Pentecostalism and the larger Charismatic movement in places such as Brazil, serve as a form of liberation theology, which, turning Marx on his head, allows followers to use religion as the material and ideological basis for social change.

Other scholars hold the slightly romantic but overly simplistic view that humans crave a system of meaning to give structure to their mundane but chaotic lives.[2] These explanations fail to explain the meaning of underground religion and cult practices for the people of New Orleans.

Rather, many urban dwellers crave self-actualization and make use of the many possibilities the urban environment offers for self-transformation and social change. They strive to refashion the world in their own image and reclaim control over the relationship between their actions and consciousness. The art of living in the postmodern urban world is an act of resistance—it's the artistic practice that combines revolution and play as well as political acts and acts of pure fantasy. This theatrical display showcases itself in the underground mystic scenes of New Orleans.

OCCULT SWAMP NIGHT IN NEW ORLEANS

The burlesque dancers appear about every thirty minutes, some naked, some with snakes, others naked with snakes.[3] A woman with white-blonde hair full of big curls and wearing a see-through leotard with small Xs over her nipples takes the stage as an audience of fifteen to twenty people cheer. A mustachioed emcee introduces her to the anything but lugubrious crowd. Ray Charles's "Hit the Road Jack" plays in the background as the dancer playfully shakes and shimmies, flirtatiously moving through the crowd. Everyone looks at her voluptuous ass as she slinks by. She jumps on the bar and, with her big beautiful brown eyes, gazes lustily into the surprised eyes of a man sipping a cocktail. Jonathan and I watch as his girlfriend Penelope shows her gratitude to the nocturnal dancing beauty by dropping a dollar in her bucket.

We're lit from hours of drinking that began the moment I picked them up from their Ninth Ward apartment. We drove to my place in Faubourg Marigny and walked up Decatur Street for an hour or so of drinking outside the bar while waiting for the show to begin. Now, suitably lubricated, we watch as another act produces a black-haired woman with short bangs and a smoking body. Penelope dances in the crowd to old-school doo-wop blaring over the speakers as Jonathan and I talk about his experiences as a leading member of an occultist group that practices Satanism.

He talks about sneaking into New Orleans cemeteries, playing in Satanist-inspired death-metal bands, and "wooing a grave": performing rituals to acquire human remains—skulls, bones, and full human heads, which he then displays on altars in his house. He relays stories about social life among occultists and the struggle for legitimacy among the community's members and various groups. It mirrors street culture in many ways, he says; questions of who is the real deal are critically important. Jonathan leans back to take a moment to think, placing his left hand over his goatee and then slicking his gold-blond hair back while glancing at one of the burlesque dancers, whose gyrations expose just a hint of her labia majora. He compares the occult to street-gang members fighting for legitimacy and social prestige—the occultist version of social capital. Just as drug dealers compete in the street, he explains, occultists compete for recognition. The question of who is legit is always important in the occult community.

Jonathan calls the various occultist groups "crews" and says he frequently initiates new members at a special location in the city. I ask if he would be willing to take me to the initiation spot, to which he enthusiastically agrees—and tonight is a perfect night for it. He dances to a few more songs with Penelope while a pretty young girl runs her hands down my chest during a lusty dance.

Eventually, we grab our drinks—and buy even more—while heading to the car. We stop at their house to change clothes and pick up some essentials, including a candle, flashlight, cigarettes and rum for ritual offerings, and a box of "cursed" items that was left on their doorstep on Penelope's birthday. The box contains several sharp metal objects and a book called *Where's Mom Now That I Need Her?*—its cover an image of a smoking frying pan on a kitchen stove. They do not know who left the box but were told by neighbors that it was a white male who sat on the porch assembling the items in the box before setting fire to their lawn using pornographic magazines as tinder. Penelope says it is especially troubling because she is not aware of anyone who would want to do this to her; it felt particularly dark and personal because her mother died when she was very young. Jonathan and Penelope are convinced this is a gris-gris, or curse, placed on them; it is time for evasive, and offensive, action.

Jonathan shows me three altars in his house, each containing an assortment of what appears to be human bones and skulls with dollar bills sticking out, semiautomatic and fully automatic weapons, bullets, cigarettes, coins and dollar bills, jewelry, bottles of rum, candles, feathers, miniature statuettes of wicked-looking characters, objects adorned with images of animals with horns, snake skins, a Saddam Hussein figurine, and other peculiar oddities and generally demonic-looking symbols. The human heads look real. He claims one of them is that of a local celebrity.[4] The three of us eat some magic mushrooms and head to the car with bottles of rum and beer in hand.

We drive to Jean Lafitte National Historical Park and Preserve in the Barataria Preserve, which offers a taste of Louisiana's wild wetlands just outside of the West Bank city of Marrero. The preserve's 23,000 acres include bayous, swamps, marshes, forests, alligators, nutrias, and over two hundred species of birds—and spiders. Huge spiders.

With music blaring, we drive over the Crescent City Connection Bridge spanning the Mississippi River. We follow a dark road until Jonathan points at a side road for parking. The road is closed to cars, but we easily walk around the barricades. After making sure our flashlights and lighters work, we walk to a small bridge over Bayou des Familles. The full moon illuminates the Louisiana night, the bayou like a painting too beautiful to be real. The frogs and alligators burp and tap and carry on; the mosquitoes and dragonflies buzz in our ears. We set the cursed box down to begin the ritual.

Jonathan lights the candle and sets it on the ledge of the bridge, then lights a cigarillo. He uses the smoke as an offering to the spirits and washes each item in the box before tossing it into the swamp below our feet. Jonathan begins with a chant while he blows out rum and smoke toward the alligators bellowing into the swamp night. Jonathan and Penelope offer prayers and empty the box of its contents, one by one, relieving themselves of a negativity that, if nothing else, weighs heavily on their minds. Finally, Penelope states that the book must burn completely, so Jonathan lights it up and leaves it in the middle of the road as we push further into the forest, heading to a sacred spot deep in the Louisiana swamp.

Jonathan keeps his flashlight on to lead the way but instructs us to keep ours off until we get off the road leading into the swamp to avoid the notice of police, who sometimes patrol the area. We walk a couple of hundred yards up the road as Jonathan and Penelope debate who should carry flashlights, discuss the danger of the trip, and wonder how long it will take before we see headlights shining at the beginning of the road near where we parked. Jonathan moves us quickly to a small path to the right, telling us to keep quiet—but also to be cautious of spiders, gators, and snakes. He shines his light on one large spider hanging off of a magnificently spun web. We walk a few feet into the swamp, ducking to avoid more impressively built webs; Penelope suddenly hides behind a tree. We all hear noises and wait in near-silence for two or three minutes, anticipating footsteps or flashlights from either the police or others walking through the swamp in the middle of the night. After a few minutes of squatting, Jonathan and I stand up and glance around the area, scoping out whoever might be on our trail. We wait, and wait, and wait some more.

A strange feeling creeps over me, starting in my stomach and ending with a light tingling on my neck. My intuition tells me to look up. Just an inch above the top of my head is an enormous spider. I immediately duck back down, my heart racing. While I might not have just cheated death, I certainly escaped the possibility of a scared and angry spider on my face—and a big red source of discomfort. Now come the police.

We see lights and hear the voices of two Louisiana State Police troopers searching the area with their flashlights. Jonathan and I walk out to greet a suspicious officer. We don't know where the other one is hiding. Officer Carolina (he later tells us that he hails from North Carolina) asks what we are doing and if we have our IDs or any weapons on us. Jonathan and I show the cop our driver's licenses. Penelope didn't bring hers. Officer Carolina asks several times, to the point of annoyance, about weapons.

Jonathan explains to Officer Carolina about the gris-gris targeting his girlfriend left on his porch. He explains matter of factly that we are occultists performing an essential undoing of the curse placed on Penelope. Officer Caroline nervously states that we can be in the park after dark but stresses that fires are not allowed. The other officer emerges from behind the trees, staring blankly at us. They walk with us back to the book and candle, mostly ashes but still smoldering, and collect them both. We tell the Louisiana State Police officers, "welcome to New Orleans," as they watch us head down the road and return to the darkness of the swamp, to continue our journey toward the initiation spot.

We stop to rest, drink from our shared bottle of rum, and laugh at the reaction of the police, who seemed nervous and unsure of themselves when talking with us, though they were also nice guys who remained professional during the entire encounter. Jonathan and Penelope realize they have left a book and other ritual items either in the car or at home. But the journey goes on. Jonathan leads us down a footpath and into an area where he warns of an abundance of snakes and spiders. If anything sends us to the hospital, Penelope says, it will be snakebites, not an alligator. Every so often, we stop to admire the beauty of the spiders, most of them female, and their impressive webs. These Louisiana swamp spiders create webs that stretch from a tree on one side of the path to a tree on the other side. They look like armored soldiers, with bright yellow and red highlights contrasting with their long, shimmering black legs. Their eyes stare at us, seemingly pleading with us to not destroy their homes while at the same time daring us to get close enough for them to attack. Many jungle insects have fallen victims to these beautiful ladies of the bayou, becoming entangled in their silk and eaten for breakfast.

Finally, we make it to the sacred initiation spot near an old oak tree. Jonathan and I push down large plants to make a five-foot path from trail to the Tree. Jonathan and his occult trainees perform many of their rituals and practices here, sometimes camping overnight. He expresses grief that the rules and policies of the park have become stricter recently, making it harder for him to conduct important activities.

Jonathan lights a candle between two large roots and shouts "*Das Matas*," which, as Penelope explains, translates to "of the woods," "of the plant," or "of the grass." According to Jonathan, the words call to the spirit of the woods. Tonight, Jonathan calls upon *matas* to remove the hex. Jonathan apologizes to *Das Matas* for not bringing the book or more offerings, expressing genuine humility. After spitting some rum out at the tree and blowing more smoke through the candle, he sits on a root as Penelope sinks to her knees and, much like a Pentecostal getting the Holy Ghost,[5] begins to cry out softly to *Das Matas*, her eyes squeezed shut. She begins to sing a sweet song loudly to the spirits. Her classically trained voice rolls through the leaves along with the wind. I kneel down to show respect and think about the swamps and wetlands of the increasingly fragile Louisiana coast. While the gods have yet to speak to me, I feel a strong presence that deserves reverence.

Jonathan asks Penelope to sing another one as she struggles to find a tune befitting the mood and energy of the moment. This is important, they instruct, explaining how singing just any old song is disrespectful. She searches her mind for something *Das Matas* might appreciate before singing a song about birds. When she finishes, Jonathan asks her to sing it again. After spending fifteen or twenty minutes here and ensuring the candle won't start any fires, we leave it in place and begin our journey back. Everyone seems satisfied, reassuring one another that the curse is buried deep in the jungle and that it will no longer have any negative effects.

I made other contacts throughout my research using the classic snowball-sampling technique: occult members would introduce me to other practitioners of the supernatural. Over time, I became well acquainted with those who occupy the nocturnal spiritual economy of the city.

THE OCCULT IN NEW ORLEANS

We meet at 7:15 p.m. at the Saturn Bar in the Bywater and order two happy-hour whiskeys to start the night. Josh is a regional representative and one of the founding masters of the fraternal organization Alombrados Oasis of Ordo Templi Orientis of New Orleans. He specializes in, among other things, the principles of Satan in Thelema—a religious and philosophical practice based on the ideas of Aleister Crowley. He graduated from the University of New Orleans in neuropsychology, studies philosophy and the general sciences, and practices Thelema, the discipline of Magick, and the method of Scientific Illuminism. Josh has an unrelenting passion for intellectual inquiry, mostly toward academic and spiritual pursuits. He's an avid practitioner of Thelema, but his knowledge goes far beyond this. He can talk for hours about dialectics, sociological concepts such as structure versus agency, and the problems of free will and determinism. He has blond hair and blue eyes and carries himself with a carefree but somehow confident demeanor. Josh's slightly graying beard gives testimony to his age—just over thirty-five—but his search for life's answers gives him a boylike charm and innocence. He combines his spiritual practices and philosophical training in his artistic pursuits as a lead vocalist (screamer) in a black-metal band. Josh works at Hex, an occult shop on Decatur Street, and helps his girlfriend Pandora run the well-known Mudlark Public Theatre, a puppet theater. Josh also worked with the Cuban-American documentarian and author Alex Mar on the film *American Mystic* and the book *Witches of America*, which describes occult practices including witchcraft, New Age spiritualism, Thelema and its "Do what thou wilt" philosophy, spiritualism, and necromancy.

The Thelemic Ordo Templi Orientis temple is hidden in plain sight in the middle of mostly residential properties in the Bywater. The temple holds regular Gnostic masses and initiations, religious rites and rituals, and classes on Thelema, magick, yoga, hermeticism, alchemy, and other such beliefs and practices. Josh gives me a tour of the temple late one night, explaining all the signs and symbols scattered throughout and showing me the books on occultism in its library. It's a large and beautiful place, complete with a backyard, kitchen, living spaces, and a temple for masses, shows, rituals, and other public and private events. We spend many nights talking about philosophy, politics, New Orleans, and the world of occultism. Although Josh does not engage in religious practices beyond the Alombrados Oasis of Ordo Templi Orientis temple, he is well aware that people practice Voodoo, magic, witchcraft, and Satanism throughout the city. Josh says, "They call New Orleans the mouth (of the Mississippi), but it's really the anus. And it gives people great magical powers."

Josh says Satanic practice is indeed widespread in New Orleans, as are many other occult practices. In this Voodoo city, magic and the supernatural are woven into the very fabric of urban social life. He explains that both above- and underground Voodoo shops exist throughout the city as well as scores of Voodoo and occult temples, altars, and shrines—from Voodoo Queen Sally Ann Glassman's shrines in New Orleans's Healing Center, to Houngan Asogwe of Haitian Vodou and Third Degree High Priest in four traditions of witchcraft, to Steven Bragg's outdoor shrine dedicated to the Mexican folk saint *La Santisima Muerte* in Mid-City. All of it is the real thing, Josh says, a blend of authentic, longstanding traditions and some New Age practices, and nothing like that depicted in the Satanic panic of the 1980s—a moral panic centered on the fear of Satanism and ritual abuse that spread throughout the United States.

RITUAL MATERIALS AND LAW ENFORCEMENT IN NEW ORLEANS

Damian remains secretive about most aspects of his life, though he still wants to give me a peek into the life and rituals of Satanists. He's been a practicing occult member for more than a decade and considers himself a Satanist and devoted follower of Aleister Crowley, the principal founder of the religious and philosophical belief system known as Thelema. Over the course of a few weeks, we frequented dozens of cemeteries, always at night, usually while everyone in our group was high on booze, beer, or mushrooms. He asked me to keep any physical descriptions of him to a minimum to protect his identity, as recent news reports make him nervous of becoming a scapegoat or the subject of a modern-day witch hunt. Recently, police have been searching for a man in the Northshore town of Slidell named Joshua Roques. Police searched his home and found evidence of grave robbing for occult "ritual materials" that included candles surrounding human and animal bones and dried blood set in a large cast-iron bowl. If captured, police plan on charging Roques with at least six charges, including interring or cremating human remains, unlawful disposal of remains, opening graves and stealing a body, mutilating and disinterring human remains, possession of cocaine, and possession of drug paraphernalia.[6] As of this writing, Roques was still on the lam.

Another recent incident that keeps people like Damian cautious occurred in Mid-City and involved bones found in the apartment of a self-proclaimed witch. Agents for the Louisiana state attorney general's office recently raided the so-called witch's house on South Solomon Street and recovered human bones and teeth.[7] State agents accuse the woman, who calls herself Darling, of taking the human remains that wash up after it rains around Holt Cemetery—a graveyard for the poor—and selling them online via Facebook. This prompted local officials to strengthen laws—specifically "Revised Statute Title 8 Cemeteries RS 8:653 Opening graves; stealing body; receiving same Opening graves; stealing body; receiving same"—that prohibit removing human remains from cemeteries.[8] The proposed state senate bill 179 plans to strengthen the original law and spell out the punishment for those who violate it.[9] This punishment includes a five-thousand-dollar fine or a year in prison for the first offence and two years in prison and a fine of up to ten thousand dollars for a second offence. Darling is also, as of this writing, on the lam.

HOPPING CEMETERY GATES

The diligent traveler does not simply visit New Orleans; he or she explores it, seeking out the invisible but authentic reality that is more real than whatever can be discerned on the surface. . . . To succeed at this quest . . . find some little-visited cemetery where voudou rites are said still to be practiced. If there is perceived to be an element of danger in such peregrinations, so much the better. The invisible world of South Louisiana will not yield up its secrets without a struggle.

—S. Frederick Starr, introduction to Lafcadio Hearn's
Inventing New Orleans

We meet at the Bywater restaurant Suis Generis on Burgundy Street and order two caipirinhas, Brazil's national cocktail, in honor of Brazilian spirits. A guy named Shaq—who is a drummer in the heavy-metal band in which Damian also plays—joins us. He's a tall, lanky fellow with long dreadlocks and a quiet, pleasant demeanor. They discuss business and the problems associated with one of their band members having been arrested and locked up. They also discuss an upcoming graduation ritual. They want to invite eight infernal spirits in what they call the Feast of Whores. This invitation has something to do with the "overcoming of revolution" and talking to wrathful deities. They also need to write the final song for their latest album and plan what Damian describes as "a ritual with the moon cycles," as if all these things are just everyday obstacles to address. "I'm a drunk," he says half-jokingly, "but I'll stop drinking for nine days to commune with these spirits." There is a general concern about the demonic energy; Shaq says, "We need a new seer. We need a chief for one position. The seer is the hardest part; it's the most dangerous part of the ritual." What they describe as a shamanic technique is a ritual that Damian says "gives me nonphysical inspiration." "Let's roll," Damian advises, "but first let's go see Little Freddie King at BJ's Lounge."

We walk out of the bar and head toward the car, where we encounter a large makeshift sidewalk altar, complete with candles, pictures, and numerous other items. A man I can barely see, sitting on his porch, tells us about Ms. Mary, the altar's dedicatee. "She was a good lady," he says repeatedly in a long trembling voice. Josh asks permission to donate money and pay his respects. The guy says, "Do whatever you want," and repeats again, while sobbing, "She was a good lady." Josh performs a ritual of some sort while chanting in Latin. This lasts for a few minutes. He stands up and waves goodbye to the man on the porch, who says, perhaps confused, "Alright man, thanks." We order more drinks to go and drive toward Little Freddie King's seventy-fifth birthday party at BJ's Lounge in the Bywater.

BJ's is jam-packed to celebrate the beloved New Orleans gutbucket gui-
tarist Fread Eugene Martin, better known as Little Freddie King. Born just
over the state line in McComb, Mississippi, he was inspired by his early experiences hanging out with his father and the likes of Fats Domino and B. B.
King on Summit Road, a dirt road lined with black-owned music venues that
ain't-that-a-shamed the nights away.[10] Like many of the characters in this
book, he hopped on a freight train one day over sixty years ago with a New
Orleans dream and Delta blues on his mind.

We dance and drink and whirl and twirl with the rowdy crowd celebrating all that is Little Freddie King and this wild thing we call the city of New
Orleans. From jazz aficionados and punk goths to heavy-metal enthusiasts
and noise-music fans, from Satanists and Voodoo practitioners to Catholic
and Protestant believers, New Orleans cruelly and lovingly embraces all types
of personalities and lifestyles. We drink gin and Abita beers, and we shake
hands with Freddie King, who is sitting proudly in a two-button blue check-
ered gingham sports jacket over an ivory waistcoat and an orange and ivory
striped shirt (with matching tie).

After about two hours, Damian tells me it's time to go. Everyone takes
mushrooms while heading out to St. Roch Cemetery, which is just a mile or
two from BJ's. We park a few blocks away to avoid drawing the attention of
cops or neighborhood residents. Damian carries a backpack full of candles,
masks, hoods, robes, incense, and other ritual materials; we walk carefully to
the tall cemetery gates. It is about 1 a.m. We take turns drinking from a bottle
of rum as we walk. One by one, we climb the fence and hop over the top,
landing on the other side next to the cemetery's chapel, where a statue of St.
Roch—the patron saint of dogs and the falsely accused—welcomes those
seeking miraculous healing, another of his specialties.

St. Roch is a peculiar cemetery, even by New Orleans standards.[11] It serves
purposes far beyond "normal" American cemeteries, which merely provide
resting grounds for the dead. In both the city's history and its folklore, St.
Roch Cemetery has always been a magical place. A Catholic priest named
Father Thevis built the shrine inside the cemetery's chapel in the 1800s, giving thanks to Saint Roch for answering his prayers to spare his congregation
from an epidemic of yellow fever. St. Roch Cemetery surrounds this gothic-
looking chapel, in which hundreds, and perhaps thousands, of believers
have deposited their prosthetics and crutches, anatomical *ex votos*, and

vintage medical artifacts—along with candles, rosaries, and other religious iconography—in a small alcove as both a sign of appreciation and a symbol of recognition that this saint has delivered on prayers asking for a miraculous healing of diseases and deformities. Many visitors leave coins, candles, rosaries, crucifixes, handwritten notes, small bottles of rum, cigarettes, pictures of loved ones, and other personal items on the altar in hopes of delivery. Even the classic book *Gumbo Ya-Ya*—a collection of Louisiana folk tales and customs—records one New Orleans woman named Luella Johnson talking about gris-gris used at St. Roch Cemetery, stating:

> If somebody treats you bad and are mean to you, git yourself some black candles and go to St. Roch's Cemetery. Light one candle before each of nine tombs, any tombs will do. When you gits to the last one, turn your back to it and hit it hard as you can and say, "Oh, Lawd! Remove this stumblin' block from my path." In nine days that man gonna die or leave you alone.[12]

Once over the gate, we head toward the chapel as Damian explains some of its history. "This was a Germanic neighborhood, and St. Roch is a patron. It's a healing chapel for the sick and it's got associations depending on your background with Legua and sometimes with Ominoo. It has associations with the Aziz spirit and the spirit of St. Lazarus. . . . It's also one of the most devastated fucking neighborhoods in all of New Orleans." I tell him, "People get popped here all the fucking time. I have a blade on me and that's about it." "We have three. Unless we run into five [people], we're OK." Damian agrees and says to Shaq, "Wait, I just want to say, Shaq, this makes me feel like the mushrooms we done before never worked. I can feel my right hand pulsing right now."

As we proceed to the chapel, Damian says, "This is the vault we wanna go in. This is the cognoscenti that has the healing chapel." We approach the altar where a statue of St. Roch stands over the tomb of the buried Jesus. We note the candles, handwritten letters, crosses, crucifixes, dollar bills, animal (or human) bones, red flowers, and coins placed at the feet of St. Roch. His statue hovers over the dark shadows of the chapel. He appears alive, ready to step off the altar to administer healing to the infirm. His tired but bold eyes, filled with sorrow and strength from years of curing sick souls, stare somehow still innocently into the beyond. He wears a gray robe and shawl, with a satchel hanging over his left side. In his left hand is a simple wooden crucifix. His right hand rests at his side. Long white hair protrudes from his gray peasant's hat, and an untrimmed beard sticks out of his golden face. A long black rosary hangs over his body. A dog stands near St. Roch's feet, looking over her shoulder toward him. The gray paint is worn from years of sweltering humidity, hurricanes, and the touches from hands of hopeful believers whispering their desperate pleas for healing.

The chapel consists of two rows of pews with about five aisles on each side. The stench of stale mold and mildew fills the chapel, from its blue ceiling to the marble floor. Various statues, religious images, paintings, and candles decorate the interior. I feel like I am in some forgotten and lonely museum filled with magic and mystery. Damian redirects our attention, stating, "So that's the Milagros chapel if you want to go take a look to the right." We look where Damian points to a black gate protecting an alcove. Its floor is made of marble bricks with the word "thanks" or the French *"merci"* etched on each. It's a small room, but it's filled with prosthetics, crutches, medical devices, and other artifacts, from pictures of saints and religious figures to peculiar dolls and handwritten notes of thanks.

Damian says, "This is a place of power." He begins what he calls the three rituals to enter the cemetery properly. These rituals involve communicating with Exus, or powerful male spirits, namely, Exu Porteira, Exu Caviera, and Exu Omulu. These spirits serve as the gatekeepers to the Reino da Kalunga, or the kingdom of the cemetery. Damian begins a sacred conjuration ritual invoking one of the chief guardian spirits, Exu Porteira, to grant safe entry into—and exit from—the cemetery. He explains the importance of respecting the powers of the cemetery, especially the gatekeeper of the dead. The second ritual involves invoking another chief guardian spirit, called Exu Caviera, who Damian says is another of the "nine chief guardian Exu spirits" found in the fourth Quimbanda Kingdom. Damian explains that Exu Caviera is a skull figure and one of the reigning entities of the cemetery. This spirit commands respect, and his permission is essential for safe entry into the graveyard. Finally, Exu Omulu heads one of the lines of a legion of spirits. Damian calls upon Exu Omulu to open the gates of death so we can speak to the other cemetery spirits. "He tells us what we need to know. We let all three spirits know that we are here in a respectful manner but also that we want the spirits to come out and play. Exu Omulu allows the spirits to talk to us, communicate, let the spirits open up." Damian offers a Latin chant to each of these spirits.

We walk through the cemetery, passing over the aboveground tombs and stumbling over a desecrated burial site, perhaps vandalized for human remains. We walk past a few concrete mausoleums before encountering one crawling with hundreds of cockroaches. It's an eerie sight; none of the others has a single roach. Further, there is no evidence of cracks or holes that could attract roaches to a corpse. We walk further down, and suddenly Damian says with urgency, "And there he is. There's a bird up on the third tomb." We see a black bird sitting on top of a mausoleum. Damian explains that this is one of the most important burial sites in the cemetery, and the bird, apparently, symbolizes the main gatekeeper of the dead. Damian offers another chant in Latin, directed at the bird. This chant, he says, requests safe passage further into the cemetery. The bird stares at us as Damian shouts toward it in Latin. After the chant, he offers thanks and tells us all to leave change, dollars, or cigarettes before moving forward. The bird is perched atop a mausoleum only a few feet from us, and it watches as we pass right next to it. The bird flies away the moment I, the last of us, pass the mausoleum. It seemed to be waiting for all of us to walk past before flying away. Noticing, Damian says, "Just give him a dollar for passage as a sign of respect. St. Roch is really well known for people coming here to do work. This is definitely a place of Voodoo—sacred land for sacred communities. Different traditions have different ways in which they enter."

Damian goes on to explain the Catholic tradition of the Stations of the Cross depicting Jesus's journey to Mount Calvary for his crucifixion. He relates the Stations of the Cross to St. Roch Cemetery. I tune in and out of this lecture, absorbed by the raw energy of the place while wondering what the hell is going on with that bird and all of those roaches on that mausoleum. Damian ends the lesson by explaining that tonight we have called upon the spirits to speak with us casually, and part of what they wanted to say involved the cemetery and the Stations of the Cross.

Leaving the cemetery also involves rituals. Damian instructs us to "leave something, like some cigarettes—you leave three coins and a few cigarettes and use some booze and basically give to respect to the Exus. You should *always* bring something. You leave the gatekeepers a keepsake to make sure that your experience while you're in here is in communion with and in harmony with the reason you entered." He tells us that failure to pay due respect entering and leaving the cemetery can have tragic or at least highly undesirable consequences. We pay our respects to the spirits, leave money and rum, hop over the gates back to the car, and head straight to the St. Roch Tavern to continue drinking. Our night's not done yet, though. We've got another tour on tap, of a secret location where Damian keeps various altars to perform occult and Satanic rituals.

SATANIC ALTARS FOR CONJURING SPIRITS

We walk into an undisclosed location where Damian and a group of fellow occultists construct altars for various religious ceremonies and rituals. These altars serve many purposes, but the main purpose is to acquire power and protection. These altars are much like the ones described earlier in the chapter and also resemble, to my eyes at least, the altars of voodoo shrines found in Priestess Miriam Chamani's Voodoo Spiritual Temple and Queen Sally Ann Glassman's shrines in the New Orleans Healing Center.

At first glance, the altar is a peculiar sight. It looks like an alien structure, something not from Earth. One glances over the details without exactly knowing what one is seeing. The feeling resembles the initial impressions of a novice when first looking at abstract art. The altar sits on a low table draped in a faded red cloth. The altar itself is tiered with shelves, each covered in an assortment of bottles, candles, and other offerings: bullets, a lucky cigarette, wooden beads, pins, and feathers. A stuffed bat peeks out from the left-hand side of a shelf, and a half-full goblet sits at the base. A human skull stuffed with and surrounded by offerings at the center of the altar serves as the focal point. Everything else seems to exist for and in service of the skull. There is money set into the eye and mouth holes, cigarettes laid out beside it, a long dagger set before it. Candles surround the skull in addition to small statues of what seem to be other, complimentary spirits. Beside the altar, next to a heavy chain that hangs from it, are more offerings laid out on the floor, including candles, long black feathers, unrecognizable metal weapons, guns, and more money and cigarettes. All of this serves to feed the spirit from this once alive and breathing human being and rescue—or drag—its spirit from the underworld, manifesting it to material reality.

Like a museum curator, Damian explains the background and cultural context for one of the altars. "This particular altar that you're taking a look at here is kind of designed after the Afro-Brazilian practices of Quimbanda. You can find things like this in Rio and São Paulo and urban places in Brazil. It's really sort of like a Congo tradition that came over with the Portuguese slave drivers. It developed down there out of Macumba, another form of spiritualism practiced in Brazil." While it is possible to locate the general context for these religious practices, he points out that there are many religious influences from which these practices derive and great debate on their origins between the practitioners. But what's most important is that these spirits come from the tough urban streets of Brazil. Damian puts it this way:

Like with any of the altars, any of the houses, and any of these practices, you're gonna have people with all kinds of fuckin' opinions about where it came from and how and why, you know, what are the politics of practicing it and this sort of thing. But practically, most of the people I talk to, including Brazilian friends, there's like street level, like every street, if you live out in the favelas, there's some mother fucker that like, is like working with these spirits to do some shit, to do something.

In essence, Damian argues that these spirits are born out of the religious practices of residents of Brazilian favelas—or slums—as they try to find, like so many others, creative solutions to the collectively experienced structural problems of their lives.[13] Damian says some residents of the favelas work with these spirits to make things happen. Going back to the favelas, Damian explains, "If you live in Brazil in a favela and the only way for you to get out the favela is to sell drugs and pimp and kill people, or at least that's what you think it is, you're gonna like end up working with demons and the demons that you create." These gangsta demons serve as infernal spirits that help residents in their journey out of the favela. Just like its residents, the gangsta demons of the favela look for redemption and a way out of the 'hood. Damian urges, "It is real important to get a grasp of why this is, and why it is scary, is that some serious power that can be tapped." As in many religious practices such as Vodou and Pentecostalism, the material world of everyday life is a direct manifestation of the supernatural world. Just like there are marginalized people, there are also marginalized spirits. And just like marginalized people searching for respect, dignity, and opportunity, these spirits search for the same qualities. Further, just as many poor residents lack the institutional resources that offer educational and economic opportunities for upward social mobility, spirits in the underworld suffer from the same type of exclusion. Instead of social exclusion, these demons experience a sort of spiritual exclusion from the forms of social capital offered in the spiritual world. Damian says to this analysis, "Yes. Exactly. What's going on here has a reflection in the underworld."

This relationship between people in the physical world and spirits in the supernatural world is not unilateral but dialectical. It's the spiritual version of people experiencing social marginalization growing up in some of the grimmest of urban conditions. What happens in the spiritual world causes the material manifestation in the physical world, which, in turn, causes the manifestation in the spiritual. That is, the physical and spiritual worlds simultaneously influence each other. The physical is both the product and producer of the spiritual world. Damian puts it this way:

> It could work the other way, and so since they're both, and when you bring it down, that's the act of magic. That's what magic is. When you look at the connection between your world and the spiritual world, but then you see a connection there that you end up bringing down to work for you. You're right. It's dialectical. They both shape each other. They're in tandem. They work with one another.

There's a practical application to magic, and the skulls and bones provide it. They harness the energy of people who've experienced anger and torment in the physical world, and that energy carries over to the supernatural. Damian creates altars with human heads and bones to pull these supernatural forces from the underworld back into the physical world, and he uses its power to gain personal strength and power and to attack and counterattack his enemies.

Damian points to one of the statues near the human skull:

> It's like the lonely soul, the soul in purgatory that has unfinished business in this world and that's connected to the skull that you're looking at. This individual [the skull] has stuff to do and sort of functions as the keeper. . . .

But in the case of the guy that you see here, I mean really, he's come back to do whatever work he's got unfinished, and in exchange we feed him, treat him well, offer him a home and better place, a holy place, rather than a place in the dirt.

Damian finds most of the skulls and bones from what he calls disrespected graves, what others refer to as a potter's field. "They were just left in a trash bag," he says, "or somewhere not properly buried. They're sort of like that malcontent or rather the disrespected dead, and so instead, we give them a place of honor."

Malcontent and tormented souls provide potentially powerful bones and skulls from which to harness spiritual energy to use for personal empowerment and as weapons to attack enemies. But this energy must be respected. A mutual relationship develops between the supernatural souls and those that use them for power. Damian explains, "There are things that we need from them, and there's things they need from us. All of this appears here." He points to the statues and other elements of the altar. "Each one of the statues is sort of like one of the natural forces, one of the personalities of spirits that comes through while we communicate with this gentleman [pointing to the skull]. He's sort of like the ambassador to the rest of this hierarchy in this realm."

Altars can serve multiple uses, from monetary gain to tools for lashing out at enemies. "Sometimes when you have people that are speaking against you and you need them to stop, you know? Other times, to secure a good family and a marriage, use this gentleman here. And then sometimes, you'll want other teachers of various forms of witchcraft and magic depending on the spirit, obviously." Each statue holds a great deal of symbolism. It seems one needs years of study to figure out the meaning of everything, much less how to practice these religious beliefs. Damian points to another statue.

Here is Aba Yoruba Orisha, who is associated with the forest and heals diseases. If you look at this statue, you can see the skull wearing a robe holding the casket of a small baby. There are many associations with yellow fever. The spirit can both cause and remove the disease. There's quite a bit that can go on here. . . . You have spirits associated with gambling, but also spirits of protection of the house and the removal of spells of people who try to work against you. You can kind of, you would ask him to help you break off whatever's been set against you. There's a lot of gambling, you'll notice. With a lot of like, luck. This is sort of an attempt to turn chance to your direction. When there's something that's uncertain, you might use that to break through that uncertainty. Otherwise, there's quite a lot of bullets on this altar and some of that is definitely vengeful where it's like protect me but protect me in, like, strike back at these assholes. And depending on what you do in your world, that may be more or less important for me.

Practicing this type of magic and witchcraft can be treacherous, especially to those inexperienced with such things. "Fucked-up shit started happening all around of us. We had all this stuff [skulls, bones, etc.] but no protections and little experience on providing offerings and taming the beast. Fucked-up shit happened, but we learned and gained control of the situation." People that become involved in these supernatural practices, especially harnessing spirits from the underworld, begin to perceive reality differently.

You just start noticing things, you become more aware. For example, the day before the car got booted, we had this guy, this old meter maid going off. He was talking about booting cars and shit and the next day the car's booted and it was like a warning. If we just like listened, "Oh, you need to watch out where you put your car," you know what I mean? Maybe it's not connected. Probably it's a coincidence. But when you practice this kind of shit, it just seems like you have those coincidences happen all the time. People who claim, "I saw that beforehand and then avoided this, and I saw what happened to the other person that didn't," and so on and so forth, and it's just kind of like, I don't really need a metaphysical model, I just know that if I can make it work for me, I can make it work for me, and I have in the past. And many, many other examples happen that keep coming up. I've been all over the world doing this, I've had many opportunities because of this type of stuff that I don't believe I'd have otherwise.

Besides the danger and the changes in one's orientation to reality that results from these practices, Damian finds his life is more fulfilling because of them. "I certainly believe my life is richer, more intense on the most basic levels, like the most existential level. There's just stuff going on and interests and creative impulses that come through that would just never otherwise happen. There is never a dull moment."

Deep in the cultural underbelly of New Orleans there exists dozens, if not hundreds, of Voodoo, occult, witchcraft, and Satanic practitioners that transcend the ordinary and carve out unique biographies in the most creative of ways through their supernatural practices. Ultimately, there is no way of confirming the intended effectiveness of such supernatural practices. Rather, all we can satisfactorily conclude is that down here in New Orleans, where transgressive practitioners of the supernatural hop over cemetery gates and take magic mushroom rides into swampy bayous, people create their own magic.

———————

Satanism and other occult practices of the supernatural abound in the above- and belowground spiritual economy of New Orleans. It's certainly nothing that poses a danger to society. Nor does it warrant the types of fears associated with the "Satanic panic" of the 1980s. New Orleans itself seems like a supernatural entity, but not some reified one detached from the subjective imprint of its inhabitants and all those that call her home. Like Damian's skulls, New Orleans harnesses the energy of those souls that once lived, and continue to live, in her magical bosom.

Perhaps the magic that persists in New Orleans will eventually succumb to the dry, sterile, and soul-crushing forces of modernity that flatten beauty and homogenize differences, but at least for now, it lives and breathes. The city gives its people the ability to project their subjective selves, their essence, onto a city fashioned from the collective will of its people. The imprint of the city reflects all the freak clowns, brass bands, jazz legends, urban black youths, buskers, street performers, cooks, ghosts of the notorious "Swamp" and "Storyville" eras, African and Caribbean slaves, Mardi Gras Indians, Voodoo queens, and even our modern occultists and Satanists that find power in this city at the mouth (or anus) of the Mississippi River. The billionaire elites and their puppet yuppies might one day take over this city, and it might happen faster than most of us want to admit. But for now, the people of New Orleans and all that call her home continue to make our own forms of magic, making this place unique in the world, a place where subcultures of transcendence continue to rise above the ordinary.

CHAPTER 8

———

GENTRIFICATION AND VIOLENT CULTURAL RESISTANCE

The stars are a free show; it don't cost anything to use your eyes.
—George Orwell, *Down and Out in Paris and London*

HI-HO, HI-HO, WHERE DID THE LOCALS GO?

Saturday nights in the deep New Orleans June summer stick to the skin like the juice of a satsuma. It's "bar-in-the-car" night as the down-and-almost-out Jim Lightfoot and I head out with a few dollars to the Rouse's supermarket in Mid-City for a bottle of NOLA rum, along with some Coke, to put in an ice-filled cooler in the back of his car. We invented the "bar-in-the-car" concept years ago to ameliorate the costs involved with drinking out in the town. By virtue of the local laws that allow for the carrying of open plastic containers of alcohol while walking on city streets—a law that routinely stuns and delights visitors—New Orleans offers a prime setting in which those low on cash can provide their own drinks between stops at local bars, pouring cheap cocktails, wine, or straight hard liquor into plastic cups from the back of their car. Most drinking venues along Decatur Street between Ursulines Street and Esplanade Avenue in the Vieux Carré allow people to enter with drinks purchased elsewhere. We still buy drinks in these bars, just fewer of them, to offset the costs of a long night of ruffian debauchery. We don't need to purchase plastic cups; every bar in New Orleans gives them away for free. "To-go" cups—"go cups" in local parlance—sit near the doorway or on the bar for customers to use when they leave. Transferring what remains of your drink from glass to plastic is part of the city's barhopping ritual. Of course, all this is only possible because it's legal to drink on the streets of New Orleans. And some New Orleanians, myself included, find it strange that people in most other U.S. cities cannot legally drink outside. Outside drinking puts more people interacting in the streets and more eyes on its public spaces. This, I argue, would make city streets even safer. Many other countries that allow people to drink outside have few, if any, problems associated with drinking on the streets.

We drink at Molly's at the Market on Decatur Street, take a whirl and twirl at the former Whirling Dervish, where blond-haired goth wannabes awkwardly dance to nouveau-punk and goth music. We reminisce about the Hi-Ho Lounge in nearby Faubourg Marigny, which once catered to all the punks, goths, anarchists, and otherwise underground culture freaks of New Orleans before Hurricane Katrina. Now to the Hi-Ho we go.

Deep funking and rare grooving to the beats of DJ Soul Sister, spinning underground disco, funk, boogie, hip-hop, and '70s jazz, happens all night long at the Hi-Ho Lounge. DJ Soul Sister, who also hosts the "Soul Power" show on New Orleans radio station WWOZ, deejayed her first live "Hustle" dance party for locals at Leo's in the Bywater back in 2004, before moving the show to Mimi's in the Marigny, and—after a noise dispute with neighborhood residents—now hosts the free event at the Hi-Ho Lounge in Faubourg Marigny. But the people don't look like locals anymore.

We walk through the door to find a crazy, packed space of mainly white faces awkwardly moving their bodies to the soulful sounds of the spinning funk. We walk through the venue, listening to the loud voices, inspecting the pale faces, watching the generic movements, sensing the lack of something otherwise known as culture, wondering, "What the fuck happened in here?" We walk to the back courtyard, where a former pop-up chef with parents from Cuba now sells Cuban-inspired food to—well, who *are* these people he's selling it to?

A timid, scrawny guy wearing a large and goofy black top hat entertains -two women. It becomes clear, after a painful ten minutes, that this poser advertises himself as a Hollywood director now in New Orleans to make it "big" while looking for local talent. One woman hails from Brooklyn; the other equally blond woman is from Maine. The Cuban chef in the back (selling what smells like damn good pork sandwiches) was born in Miami, and others sitting at tables and smoking cigarettes have arrived from Milwaukee, Mississippi, California, New York, Ohio, and Montana.

Candice, the big and sweet but ornery black woman outside checking IDs, maintains a blasé countenance with every patron entering the venue, no matter what clever thing they try to say. She's from New Orleans East, just east of Gentilly. A local girl. Good. The IDs read: Wisconsin, Wisconsin, California, Rhode Island, Massachusetts, Alabama, Alabama, a bunch from Mississippi, Missouri, Missouri, Illinois, Michigan, Michigan, god knows how many from Texas, two passports from Great Britain, four from Iowa, then Virginia, Virginia, and five or six New Yorks.

Then: A Louisiana driver's license! Finally. "Yo," I say, "you from here, bae?" She asks, "What?" then replies, "No, I moved from New Jersey." Candice from New Orleans East says, "None of these people from here. They all from somewhere else, though some of them live here now." After an hour of looking at IDs with Candice and asking those with Louisiana IDs where they're from, it becomes clear, as Candice said, that none of them hail from here.

Inside, we watch as two young guys, ringed by an impressed audience, compete in a sort of robot dance. We've found them! Locals entertaining tourists! Close, but no. They've come from the Gulf Coast of Mississippi and Alabama. Moving through the crowd, we bounce to the beats, smile at the frenzied dancing crowd digging the Soul Sister's mix. "You from here?" "No," "no," and "no." Everyone says "no."

It's a gentrification station here at the Hi-Ho, where the locals once lounged. And where did the locals go? We just don't know. "Do you know what it means to miss New Orleans?"—even when you're in it?

FUCK YUPPIES

"Fuck yuppies." "Yuppies = bad."

The graffiti shocks the multicultural neoliberal customers expecting to get some crackerjacked café lattes and muffins at the newly opened St. Roch Market in the Bywater on St. Claude Avenue before heading into the office.

They stare with confusion at the words spray-painted on the wall. Video surveillance captured three men and two women at about 2:25 a.m. splattering pink paint on the front and sides of the building, writing antigentrification graffiti, scribbling an anarchy sign, and smashing the market's windows. The result was a brilliant display of the words "Fuck Yuppies" and "Yuppie = bad" showcased on the building's façade. Its obvious illegality notwithstanding, this act of vandalism reflects some of the tensions here in New Orleans over the hypergentrification happening in many historic neighborhoods of the city. Lacking political and economic power, some residents use such acts of vandalism as weapons of the weak to express their anger at a market that symbolizes gentrification.

The St. Roch Market has a long history in the neighborhood once called Faubourg Franklin, a place that before the Civil War was home to one of the largest populations of free blacks and later was where people such as the early jazz great Jelly Roll Morton roamed.[1] Originally built in 1875, the building was a seafood market and later a seafood restaurant. It was abandoned in the aftermath of Katrina and recently reopened after a multi-million-dollar renovation project, creating the 8,600-square-foot space, which now houses thirteen food vendors. Although its co-owner, the Alabama native Will Donaldson, claims, "This building is supposed to be about community,"[2] the specialty foods and dishes, which often appropriate New Orleans slogans—such as "Japchalaya" and "Who Dat crêpe," as well as other foods like charcuterie and cold-pressed juices—sell at relatively high prices that exclude most of the working-class residents of the surrounding communities. While City Councilwoman LaToya Cantrell publicly voiced disapproval of the vandalism, she also admitted to having mixed feelings about a market that was not built for residents of the neighborhood, given that 40 percent of the people nearby live below the poverty line. "The market is great, but when I went there it was bittersweet, because I could feel that the community could see it as: 'This is great but it wasn't created for me,'" she was quoted as saying in the *Times Picayune*. "People need to see themselves in the transformation of their neighborhood."[3] This form of reified urbanism excludes the majority of the community residents from their own neighborhood establishments. Instead, St. Roch Market caters to a largely white, middle- and upper-middle-class clientele with disposable income. Walking through St. Roch Market, it feels like anything but a place to "make groceries," as the locals put it. Rather, it's more like the Chelsea Market, the upscale bar and restaurant yuppie food-carnival complex in Manhattan's posh Chelsea neighborhood. During the day in this majority-black neighborhood, one can see white middle-class people eating chocolate puffs, tartar and carpaccio, açai bowls, sipping wine and specialty cocktails like bees' knees (gin, honey, lemon, and sage). Sometimes, members of the black community can't help but stand on their outside porches to watch the unfamiliar sight of white yuppie outsiders taking yoga classes in plain view, raising their asses in the "downward dog" pose, right on the patio behind St. Roch Market.

Posh bar and restaurant spaces like the St. Roch Market, located in largely poor working- and underclass communities, intensify the feelings of relative deprivation—or the subjective experience of social injustice and inequality—for many of the neighborhood residents. It's one thing to be poor among the poor; it's another thing to experience poverty surrounded by wealth. Even other business owners in the area, like Fatma Aydin of the Fatoush Mediterranean Restaurant, express the frustration that other members of the community harbor: "We thought the city was doing something good. We never thought the city would screw us like this."[4] Well aware that communities respond to feelings of betrayal and social injustice, Lisa Suarez—who serves as president of the Faubourg Marigny Improvement Association—adds, "It's not what was promised. It was promised to have reasonably priced groceries for residents. . . . And if people feel disenfranchised, I think what you get is graffiti."[5]

Pastor J. B. Watkins of the St. Roch Community Church fears that the neighborhood will soon transition from a black, working-class community to a majority-white middle-class neighborhood of young professionals. But fear often turns to anger. That's why Pastor Watkins says the vandalism is unsurprising: "It wasn't a surprise to me that something like that happened. Obviously, I wouldn't sign off on anything like that. . . . But in light of the market contributing to some of the frustrations regarding the identity of the neighborhood, I wasn't surprised at all."[6] Leesha Freeland, the president-elect of the New Orleans Metropolitan Association of Realtors, says properties in the St. Roch neighborhood that she could not give away in the past are now selling fast. She recognizes what will happen to the longtime community residents: "It's sad because these are people who have lived there their whole lives, and now it's going to become an area that they can't afford to live in anymore. And so they're going to have to move from their homes." On the other hand, she celebrates this "exciting" transition to an "artistic, funky vibe"—while admitting that it's causing "a bit of an unusual turf fight."[7] This turf fight pits real-estate developers and middle-class white residents against poor and mostly black New Orleans residents. This "turf fight" has been going on for decades and is still in progress. But some people don't see it as a fight at all. Rather, they believe that white people enhance black places.

Some supporters of gentrification—those that suffer from "white-savior syndrome," white people who think they can "save" poor black people—believe that their branching out into poor black neighborhoods will make those neighborhoods better.[8] This is reminiscent of the author and social activist Toni Bambara's short story "The Lesson," about a well-intentioned schoolteacher sending her poor black students from Harlem to Manhattan's expensive FAO Schwartz toy store. Although the teacher's intention was to expose the neighborhood kids to the world outside of their oppressed community, the children instead experienced relative deprivation for the first time, realizing that many of the toys that white children have access to cost more than their families could ever afford. In the story, the teacher asks the students if they learned anything from the experience. Sugar, the young black woman telling the story, replies: "This is not much of a democracy if you ask me. Equal chance to pursue happiness means equal crack at the dough, don't it?" Instead of submitting to her own structural subordination and economic exclusion, she reveals her determination and agency with the statement, "Ain't nobody gonna beat me at nothin." Except that this is not FAO Schwartz. This is New Orleans, and the city can become violent when residents feel themselves under siege.

But what is the cause of these feelings? What is increasing the sense of relative deprivation? What does this have to do with the changing demographics of the city, demographics changing in part by accelerating processes of gentrification? And what might this have to do with the notorious violence of the city?

FROM CHOCOLATE CITY TO BUNNY BREAD

We ask black people: It's time. It's time for us to come together. It's time for us to rebuild a New Orleans, the one that should be a chocolate New

Orleans. And I don't care what people are saying Uptown or wherever they are. This city will be chocolate at the end of the day. This city will be a majority African-American city. It's the way God wants it to be. You can't have New Orleans no other way; it wouldn't be New Orleans.

—Former New Orleans mayor Ray Nagin, City Hall, January 17, 2006

The population of New Orleans dropped from 484,674 people at the beginning of the twenty-first century to an estimated 230,172 (July 2006) after 2005's Hurricane Katrina, a decrease of 254,502 (about a 47 percent decline).[9] In the year 2000, the black population of New Orleans was 67.2 percent; the minority white population was 28 percent, a nearly three-to-one ratio. Ten years later in 2010, the New Orleans population reached 343,829, 60.2 percent black and 33 percent white. Between 2000 and 2010, New Orleans still has a population 29 percent lower than its pre-Katrina numbers, with a 36.5 percent decrease in the black population and 16.5 percent decrease in the white population. The population has continued to grow steadily at a rate between 1.1 and 1.4 percent annually, reaching 369,250 (or 76 percent of the 2000 population) in 2012 and an estimated 384,320 in 2014. Besides higher birth rates than death rates, much of this growth is the result of people moving to New Orleans from other parts of the United States. The U.S. Census Bureau estimates that the 2013 population has 99,650 fewer black people living in New Orleans than the city's population in 2000, while the white population has decreased by 11,494. Meanwhile, in part because of the work needed to rebuild the city, the number of Hispanics in New Orleans has grown by 6,023. Though their numbers have dwindled, black people still represent the largest ethnic group in the city, at 59 percent of the population; white people make up an estimated 31 percent.

New Orleans serves as an example of the collapse of first- and third-world dichotomies inasmuch as the first world exists in the third and the third exists in the first. We now find many third-world conditions in some of the wealthiest countries, including, and perhaps especially, the United States. While 24.5 percent of people living in New Orleans lived below the poverty level in the year before the 2005 storm, the U.S. Census estimates that in 2013, 28.1 percent of the New Orleans population lived below the poverty level and that 41.5 percent of people below the age of eighteen lived below poverty. Today, poverty disproportionately affects the black population. An estimated 36.3 percent of black New Orleans residents live in poverty, while 14 percent of white residents live below the poverty line (compared to 2004, when 30 percent of black New Orleanians and 13.6 percent of white New Orleanians lived in poverty). Nearly 50 percent of the black population of New Orleans under the age of eighteen lives below the poverty line. That's half of all black children living in poverty.

Many people in New Orleans, especially members of the black population, live in social conditions similar to that of the third world, including in major social indicators such as health, literacy, and life expectancy.[10] Some scholars show how capital investment is used to bring a type of gentrification and revitalization that benefits a few members of the city at the expense of further marginalizing a large part of the population. This gentrification process has led to an increase in poverty since Hurricane Katrina, which some call a national disgrace—the "Third Worldization" of a major U.S. city.[11] For example, the U.S. Department of Education reveals that about 18 percent of people in New Orleans lack rudimentary reading skills, while 44 percent make it through life with minimal reading skills. The professor of urban studies David Gladstone, from the University of New Orleans, points out that if New Orleans were a separate country, its literacy rate would rank 102 out of 151 countries, lower than "Vanuatu, Syria, Cape Verde, Botswana, and El Salvador." New Orleans fares no better with its infant-mortality rate, ranking low when compared to the rest of the country. For black people, the infant-mortality rate reached 8.9 infant deaths per thousand live births in 2009. Again, Gladstone shows that if New Orleans were a separate country, the infant-mortality rate for the black population would rank 61 out of 226 countries—below Cuba, Antigua and Barbuda, and Serbia.[12] Few people, I hope, would disagree that these ratings are indeed a national disgrace.

Immediately following Hurricane Katrina, the U.S. military entered the city as an occupying army, toting guns and flexing military muscle, rather than as representatives of a government concerned with the interests of its people.[13] Responding to exaggerated and racist media representations of young black males robbing people and looting stores, Louisiana's governor Kathleen Blanco threatened to use the military to kill people, and indirectly—through use of such code words as "hoodlums"—she threatened to have them kill *black* people:

These troops are fresh back from Iraq, well-trained, experienced, battle-tested, and under my orders to restore order in the streets. They have

M-16s, and they're locked and loaded. I have one message for these hoodlums: These troops know how to shoot and kill, and they are more than
willing to do so if necessary. And I expect they will.[14]

Then there's the real threat of incarceration, which has long been a part of
the everyday existence of the black population of New Orleans. Oftentimes,
when authorities are not threatening military aggression to kill American
citizens (as they did during Katrina), they're just jailing them. The U.S. Justice Department reports that in 2005 the United States had the highest proportion of its population in prison in the world, at 738 people sitting in prison
per 100,000.[15] In this same year, the imprisonment rate in Louisiana towered
above the national average, with 1,138 persons per 100,000 in prison. Prior to
the storm, the imprisonment rate for black males in Louisiana was more than
double Louisiana's average, at 2,452 per 100,000 persons in 2005, or over five
times the rate of white males.[16]

Today, the U.S. prison rate continues to be the highest in the world, with
716 people out of 100,000 of the national population locked up, and Louisiana continues to hold the dubious distinction of locking up a higher proportion of its people than anywhere else in the world. A recent report shows that
1,341 people per 100,000 in Louisiana were incarcerated in 2013, making the
state the world's prison capital.[17] That's exactly the title of a recent article from
the *Times Picayune* reporting that Louisiana's incarceration rate is nearly
double the national average, with one out of every eighty-six adults in prison.
They also report that this number is disproportionately high when it comes
to black men, especially black men from New Orleans, where one out of
every fourteen remains behind bars, and a shocking one in seven are in
prison, on parole, or probation. At the time of this writing, about five thousand black men from New Orleans are in state prison, compared to four hundred white men from the city.[18] Suffice it to say, spending time in jail or
prison is an experience many black people from New Orleans know well.

As if all of that isn't troubling enough, when black people from New Orleans
are not under military threat from their elected officials during national
disasters, or serving time in a punitive criminal-justice system that targets
the most vulnerable populations, or dealing with third-world conditions in
concentrated and segregated areas of poverty, they are often finding their
communities disrupted through gentrification and urban-redevelopment

projects undertaken by the political and economic elites with big bucks to spend and even more money to make.

THE PROCESS OF GENTRIFICATION IN NEW ORLEANS

Gentrification refers to the conversion of socially marginal and working-class areas of the central city to middle-class residential use, often funded by private-market investment capital in downtown districts of major urban centers.[19] In this process, middle- and upper-class people take up residence in traditionally working-class and underclass city neighborhoods, especially historic ones. The term "gentrification" captures the class inequalities and injustices that exist in capitalist urban policies and land markets. The Bywater, or the "Williamsburg, Brooklyn, of the South," serves as a quintessential example of gentrification. Although the sociologist Ruth Glass coined the term "gentrification" in *London: Aspects of Change*, observing that the "gentry" class "uplifts" residential areas and displaces original working-class occupiers, gentrification existed in New Orleans long before the birth of this concept. The Tulane geographer Richard Campanella aptly covers the history of gentrification in New Orleans,[20] noting that writers and artists in the 1920s and 1930s, many of them mentioned in chapter 1, were the first to settle and gentrify the Vieux Carré. Part of their attraction to that neighborhood was the area's long history as well as the cheap rent and free-flowing alcohol during Prohibition. Wealthy middle-class whites displaced the blue-collar Italian and Creole locals. The wave of gentrification continued to spread east of the Vieux Carré to Faubourg Marigny just downriver. Many of these gentrifiers in the Marigny were gay waiters and other tourism workers as well as workers in upscale Vieux Carré restaurants.[21] Campanella notes that though gentrification happened early in New Orleans, its progress was slowed by the many social problems and economic limitations of the city. This all changed after Hurricane Katrina. Campanella puts it:

Everything changed after August–September 2005, when the Hurricane Katrina deluge, amid all the tragedy, unexpectedly positioned New Orleans as a cause célèbre for a generation of idealistic millennials. A few thousand urbanists, environmentalists, and social workers—we called them "the brain gain"; they called themselves YURPS, or Young Urban Rebuilding Professionals—took leave from their graduate studies and nascent careers and headed South to be a part of something important.

Thousands of newcomers transplanted to the city. Many of them possessed specialty skills in things like "new venture development," such as St. Roch Market's co-owner Donaldson, or, as Campanella points out, as new-media entrepreneurs or Teach for America teachers. Other transplants coming in droves to the city since the hurricane are artists, political radicals, poets, musicians, and other types of creatives looking for "it" in the "undiscovered bohemia" amid the history and mystery of the Faubourgs. Campanella argues that historic neighborhoods and neighborhoods in close physical proximity to already gentrified ones pave the way for wealthy white people to displace working-class residents. Certainly, Faubourg Marigny and the Bywater, both immediately downriver from the long-gentrified Vieux Carré, are now almost fully gentrified. In fact, new multi-billion-dollar urban development projects in the Bywater are installing tall building complexes that will inevitably block the refurbished older houses from views of the river. The new residents will be replacing wealthy gentrifiers with even wealthier ones. In other words, the extreme yuppies threaten to replace the moderately rich yuppies. According to Campanella's theory of proximity, Tremé and the Seventh Ward are next in line to gentrify.

Campanella identifies the four frontier cohorts, or four-phase cycle, of gentrification: gutter punks, hipsters, bourgeois bohemians, and the bona fide gentry. Gutter punks, or transient societal norm-rejecting squatters who beg for money, begin the process of gentrification. Hipsters, or educated and often privileged "obsessively self-aware" critics of mainstream society who often busk for money or work part-time jobs, follow the gutter punks and give a neighborhood a "cool" or "funky" vibe. Bourgeois bohemians, or moneyed and well-educated, usually liberal professionals who sympathize with hipsters but negotiate with mainstream society—and who become involved in urban civic life and shop at places like St. Roch Market—move into neighborhoods following the hipsters and begin to gentrify an area seriously. The last group, the bona fide gentry, who are wealthy and committed—even if liberal—to the conventional norms of society and its dominant institutions, fully gentrify a neighborhood, raising property values and displacing all or

most of the working-class and underclass residents. As Campanella estimates, "St. Roch is currently between phases 1 and 2; . . . Bywater is swiftly
moving from 2 to 3 to 4; Marigny is nearing 4; and the French Quarter is
post-4." I would add that Tremé and the Seventh Ward appear poised to be
the next parts of town to gentrify and currently sit between phases 1 and 2.

Looking specifically at the Bywater, St. Roch Market only adds to the
already established

> ten retro-chic foodie/locovore-type restaurants, two new art-loft colonies,
> guerrilla galleries and performance spaces on grungy St. Claude Avenue, a
> "healing center" affiliated with Kabacoff (a local developer specializing in
> historic restoration) and his Maine-born voodoo-priestess partner, yoga
> studios, a vinyl records store, and a smattering of coffee shops where one
> can overhear conversations about bioswales, tactical urbanism, the klezmer
> music scene, and every conceivable permutation of "sustainability" and
> "resilience."[22]

Many of these transplants appropriate the language of the local culture, like
the New Orleans Food Co-op in the Healing Center, which boasts a sign stating: "Make Groceries—Make Community." Perhaps many of these gentrifiers are really well intentioned and love the culture of the city. Perhaps they'll
love it to death.

POST-KATRINA GENTRIFICATION

Professor David Gladstone, who has lived in New Orleans for nearly two
decades, aptly describes transplants as literally "loving the culture to death." He
argues that in New Orleans, "gentrification" takes on a clear class and racial
dimension, where previously poor and black neighborhoods transition to
homogenous, largely white communities. This type of gentrification increases
segregation and decreases the cultural diversity of New Orleans. Instead of
looking at the specific geography necessary for gentrification, Gladstone
examines the links between tourism and gentrification in New Orleans.
Some of the first neighborhoods to gentrify in New Orleans in the 1970s
were the neighborhoods close to the tourist zones of the French Quarter,

184 Gentrification and Violent Cultural Resistance

where many gay waiters and other Vieux Carré restaurant workers moved, just as aspiring artists were the first gentrifiers of SOHO and Chelsea in Manhattan. Eventually, wealthier gentrifiers, like those with economic capital that control so-called art districts and urban-redevelopment projects, replaced the original gentrifiers, who could no longer afford rents and the high price of housing. Just as high rents have driven out many artists from much of New York City, few tourist workers could afford the high rents of the Vieux Carré, Faubourg Marigny, and the Bywater. Gladstone points out that Lower Manhattan, once a place for aspiring artists and creative types, now is under the private control of art-gallery owners, leading sociologists such as Sharon Zukin to ask who now controls the production of culture in the city. Similarly, workers in the tourist industry of New Orleans can no longer afford the high rents of the surrounding Faubourgs closest to the city center as a wealthy land- and business-owning class takes over these neighborhoods. If tourist workers at one time paved the way toward gentrification, today they are much less of a factor. For what Gladstone calls straightforward reasons, most receive poverty-level wages. He shows that all the arrows pointing to the cause of gentrification is the growth of tourism and the effect it has on inner-city land values through large urban-redevelopment projects that expand the city's tourist zone, including the "assemblage of hotels, conference centers, sports stadia, and historic neighborhoods."[23]

Everyone in New Orleans knows that the culture of the city is produced in the working-class and underclass areas of the city and sold in the Vieux Carré. So what happens when the wealthy, profit-driven capitalists, real-estate developers, and urban-redevelopment planners colonize the black and poor neighborhoods where New Orleans culture is produced? This is how they destroy that culture. Hipster transplants "loving the culture to death" pave the way for gentrification, then the wealthy political and economic elites of New Orleans appropriate that culture, stealing it from the culturally thick and creative streets of poor black people and selling it at a profit. The process of gentrification is a full-on economic assault that displaces poor and working-class black residents and destabilizes their family, work, community, and interpersonal relationships. Considering how many poor black residents in New Orleans find themselves excluded from resources and access to institutional support, and given the failed institutions that attempt to ameliorate conditions of poverty, poor black residents depend heavily on their community and family support networks. In fact, these support networks build resilient internal cultures within the neighborhoods that allow community residents to confront the challenges of everyday living. Destabilizing them is an act of economic warfare. If gentrification is the new segregation in New Orleans,[24] it is also a merciless act in a postcolonial age that uses neoliberal urban policies to conquer and destroy communities and that disproportionately affects the most vulnerable populations of the city. As the sociology professor John Arena argues, the once-powerful grassroots organizations that defended the New Orleans black community—by, for example, saving many of the city's housing projects—have been co-opted by the political and economic elite, using the thin veil of neoliberal economic policies, to align the city's policy makers, nonprofit organizations, community activists, and real-estate investment teams into a "foundation-funded nonprofit complex."[25] The Black Urban Regime Coalition of New Orleans works to support the largely white economic elites of the city,[26] often at the expense of their supporters, and this puts the black community in the path of an avalanche of change threatening to crush their neighborhoods.

The poor black community is now more vulnerable than ever, but vulnerability does not equal weakness. Violent crime in the city has increased over the past year. In 2015, with a total population of about 390,000 residents, the city's official crime-data report revealed that New Orleans experienced 164 murders, 409 rapes, 1,085 armed robberies, 412 simple robberies, and 1,906 assaults.[27] Many of these crimes occurred in heavily gentrifying areas such as the Seventh and Ninth wards. Violent crime, especially directed toward the newly arriving white transplant population, has increased. Perhaps this is because white and generally unarmed outsiders move into historically black areas and stick out like sore thumbs or because, as the LSU public-health criminologist Peter Scharf explains: "Gentrified folks are much less likely to shoot you dead if you rob them."[28] Since Hurricane Katrina, New Orleans has become poorer, more unequal and racially segregated, and more dependent on a tourist industry that offers only poverty-level wages.[29] As the wealthy appropriate the tourist industry, which is based on a culture that gentrification destroys, many New Orleans natives who depend on tourism to eke out a minimal existence will lose the very culture that produces that industry. In short, gentrification pushes the producers of New Orleans culture out of the historically black neighborhoods where that culture is produced. As the director of the Fair Housing Action Center Cashauna Hill puts it, "New Orleans was built on the culture bearers and musicians that made this city special. If we want those traditions to continue, then families must have access to housing that's affordable in all of our neighborhoods."[30] Although some black residents of these historic communities, especially the homeowners, might think well of gentrification, since it provides resources to the community (perhaps more evidence of racist city policies), if the current process of gentrification continues, it will destroy both these historically black neighborhoods and further destabilize the world of work, family, community, and interpersonal relationships within them.[31] Further, the very culture that fuels the tourist industry will be pushed out of the city, and this could have disastrous consequences for both the cultural bearers and the political and economic elites that profit from them. One of these rich and historically culturally diverse neighborhoods is the Seventh Ward, which is just beginning to gentrify.

White people? Oh, shit. There goes the neighborhood.

GETTING THE OKEY DOKEY: GENTRIFICATION IN THE SEVENTH WARD

At a community meeting on May 27, 2015, at Uptown's Mintz Center on Broadway Street called "Neighborhood Revival or Faux Bourgs? Gentrification in NOLA," four panelists discuss the changes going on in New Orleans as a result of gentrification. The conversation focuses on the changing demographics of the city's neighborhoods, the new arrivals, and the rising prices of rents and houses. The panelists and community members in attendance wonder about the effects these changes will have on the city, especially for many community residents who can no longer afford the high rents. New Orleans City Councilwoman LaToya Cantrell points out:

The city of New Orleans, if we just look over time in 1982, 12 percent of people's income was going towards rent. But now it's at 35 percent of poor people's income is going towards rent. Rents are through the roof, and at one point, you know, this was a place you could live in your community. You could afford to work within the hospitality industry even, which drives our economy here. But now, our cultural bearers, as I call our musicians, our chefs, our hospitality employees, are being priced out of communities that they've been able to live in literally for generations.

The topic now turns to the Seventh Ward, a community steeped in New Orleans history and an important area for the production of New Orleans culture. Vance Vaucresson, a sausage maker by trade and codeveloper of a residential housing project, used to live in the Seventh Ward prior to Hurricane Katrina and openly admits to being frustrated with the lack of investment in the area, especially compared with other areas of the city that received funding from development. Mr. Vaucresson explains his connection to the community: "The Seventh Ward of New Orleans [is] a very rich and very special place in my life span, as to the cultural significance of the people of heritage that I belong to—the Creole culture. My family has existed and done commerce and business in the Seventh Ward for over three generations." He is a student of the culture and aware of its place in the production of a distinct New Orleans sensibility. His attempts to move back after Hurricane Katrina forced him out of the area have proven just as frustrating as the city's lack of investment prior to the storm. He explains how the Seventh Ward contributed to the history of New Orleans, especially in shaping its literature, music, and politics. He now hopes, though with great skepticism, that the area will somehow find some type of investment and growth in such a historically and culturally important area.

People started gradually to move back to the area in the years following the hurricane, though the people had different faces. "What I saw was a community that was coming back. Slowly but surely, but it wasn't those that had lived there for so long. It was new people." This new and energized growth—slow and gradual—offered hope for a new vision that would take the community past its more violent "Seventh Ward Hard Head" gang days.[32]

At one time, as Mr. Vaucresson points out, the Seventh Ward served as home to a rich, vibrant, and stable multiethnic community of Muslims, Asians, Italians, and Creoles. He cautions people attending the community meeting that to make positive change, it's important to understand the culture of the community into which you are moving. "You only know where we're going by knowing where we come from. And it's [knowing] the particular part of that culture, the people, and it's amazing." Specifically addressing the members of the audience that are transplants, he says directly, "I don't know how many of you are gentries [gentrifiers] . . . in the Seventh Ward, [but] we have people who are trying to just deal with not only who we're living with but our own identity because in the history line, we were [historically] dealing with a different caste system than a lot of other places were dealing with in the country." Mr. Vaucresson goes on in depth about the conflicts that emerged in the Seventh Ward between French Creoles and white residents as well as between Creoles of different colors. Many of these conflicts emerged because of racist political and economic policies—from Jim Crow to urban-planning policies that placed a stretch of Interstate 10 on Claiborne Avenue, cutting directly through the neighborhood. Mr. Vaucresson passionately states: "Five white men made a decision to not put the interstate on the river. 'Let's put it down Claiborne and destroy all of those wonderful oaks.' And then they destroyed a neighborhood, and they left it to rot." This has left the residents of the community skeptical of the political and economic elites, who have historically created policies that have negatively affected the community, as well as of the outsiders moving to places like the Seventh Ward and who perhaps unintentionally begin the process of gentrification. Now that white outsiders are moving into the Seventh Ward, plans for redevelopment have already begun. As Mr. Vaucresson explains, "All of a sudden, post-Katrina, we get the Livable Claiborne Corridor study."

According to the *New Orleans Tribune*, the Livable Claiborne Communities Initiative is "a comprehensive and strategic study that focuses on the economic development and community revitalization of the Claiborne Avenue Corridor, specifically the elevated section of Interstate 10 between Napoleon and Elysian Fields." I-10 is an urban highway viaduct designed in the 1960s and placed through the center of New Orleans. It runs through Faubourg Tremé and the Seventh Ward. According to the *Times Picayune*, the controversial interstate along Claiborne Avenue—formerly known as the "black people's Canal Street"[33]—led to the decline of the vibrant and relatively prosperous Tremé and Seventh Ward neighborhoods. With the help of Robert Moses, the same redevelopment planner who isolated and destroyed the Brooklyn neighborhood of Red Hook with the routing of the Gowanus Expressway, the original plan was to put a Riverfront Expressway through the Vieux Carré.[34] French Quarter activists with political and economic clout pushed the project to the poorer, and blacker, neighborhoods along

Claiborne Avenue. The plan involved clearing out oak trees on the wide Claiborne Avenue median, small community businesses—from Joe Sheep's sandwich shop to Moe's pie shop—and about five hundred residential homes of largely black and disenfranchised residents.[35] The construction of the elevated I-10 expressway proved to be "the most visible and painful blow to commercial and residential life on North Claiborne Avenue."

This blow perhaps most affected the once-vibrant Faubourg Tremé neighborhood, which borders the Vieux Carré. John Hankins, the executive director of the New Orleans African American Museum, describes Tremé:

> Tremé is a magical neighborhood; it is in fact the oldest continuously inhabited African-American neighborhood in the country. It's the oldest neighborhood of free people of African descent in America. I'd like to think that if the French Quarter is the heart of New Orleans, then Tremé is certainly the blood that flows through it and gives it its energy and its soul. And Claiborne Avenue was the centerpiece of that culture of New Orleans that makes her so unique that people come to New Orleans to experience. The culture used the Claiborne Avenue strip as its lifeblood. That's where people every single weekend would fill the neutral ground as we call it. They lived that culture on those grounds every day, and certainly on weekends.[36]

Although Tremé was gentrifying prior to Hurricane Katrina, the process has dramatically increased since the storm. Some of the neighborhood is now predominantly white, and if the trend continues, it will become fully bleached of its black residents.[37] As Gladstone grimly puts it: "The oldest African-American neighborhood in the United States—a neighborhood whose residents have historically been major contributors to the production of the city's distinctive cultural forms—will soon have relatively few black people living in it." Gentrification in New Orleans has once again put a wrecking ball right in the backyards of the black community.

Now, as the city's political and economic elites entertain the idea of removing the Claiborne Expressway, community residents are understandably wary of city planners who say they intend to revitalize their community with big urban-renewal projects. Although the Livable Claiborne Communities Initiative promises "economic innovation and inclusion, and access to cultural, employment, transportation and housing resources," many long-term residents believe that city planners will remove the Claiborne overpass to gentrify the area for economic gain and at the expense of the current residents. As the *New Orleans Tribune* states, believing that this urban-revitalization project is an attempt to push out longtime residents is "not a far stretch when one considers that the proposed redevelopment of the Iberville/Tremé area (including the redevelopment of the Iberville Housing Development) under the Choice Neighborhoods project that seems to intersect with the Livable Communities study along the stretch of the overpass between Tulane and St. Bernard avenues."[38] As one John Hankins puts it:

> Now we face a situation where once again, this area is being considered for another major transformation if the I-10 is taken down. And the area becomes much more attractive to developers and people who are moving back in to the city, then many people fear that there will be a gentrification and a consequence of that and that gentrification would make this place almost too expensive for the people who have lived here and whose families have lived here for so long to remain.[39]

Perhaps the economic and political elites of New Orleans do not intend to displace poorer black residents. Maybe they care about the communities their economic plans disrupt. The website NOLA Anarcha describes the well-known New Orleans urban real-estate developer Pres Kabacoff as "a piece of shit millionaire developer" and "destroyer of the St. Thomas Projects, builder of hideous condos, Wal-Mart partner, and general rich bastard." When asked if the poorest were just "screwed" in the gentrification process, Pres Kabacoff, the Robert Moses of present-day New Orleans, responds in the following way:

> In terms of race, black people in this town have less money. When neighborhoods revitalize, I think it chases all the poor out, and in our city the poor are almost all black, so it's more a coincidence. And there is probably some racism involved in that. That's the downside of neighborhood improvement. But if the solution is don't allow people to sell or improve their houses, then you haven't done that neighborhood any good.[40]

New Orleans culture both embraces and resists change. When such change appears inauthentic, all bets are off. Almost everyone talks about gentrification in New Orleans. It's one of the hottest and most controversial topics in the city. Many longtime residents of New Orleans have seen the monster of gentrification and know its lies. Going back to Mr. Vaucresson, he explains, "You can't think that everybody is going to fall for the 'okey dokey.' You can't think that we don't see how the area's being tendered for development. . . . I'm seeing the beginning of gentrification. I'm seeing the bohemians and the artists move in and settle among the people." It's not just Mr. Vaucresson who refuses to accept the "okey dokey." The residents of gentrifying neighborhoods like the Seventh Ward also refuse to trust the originators and planners of gentrification. Some people develop their own cultural responses to what they perceive as threats to their communities, and sometimes those responses can be deadly.

VIOLENT CULTURAL RESPONSES TO GENTRIFICATION

Black New Orleanians have faced a military threat from their own city leaders during times of crisis. They endure poverty and social conditions that compare to the third world and global south, face a disproportionate arrest and imprisonment rate in the United States (and perhaps world) because of a punitive criminal-justice system that targets the poor, and—on top of it all—must deal with gentrification and redevelopment efforts that wreak havoc on their family, community, culture, and interpersonal lives. Although many black middle-class homeowners welcome gentrification into their neighborhoods because it brings increased resources into the community,[41] many renters and poorer community residents realize that gentrification pushes them out of their neighborhoods. As explained above, the few black elites, as Arena argues, borrowing from what Adolph Reed calls the Black Urban Regime Coalition, that have defended the black community in New Orleans in the past have been co-opted into the very dominant culture that undermines the community.[42] Now, as middle-class and relatively wealthy white people move into the areas around St. Claude Avenue and the Seventh Ward, some in the black community are developing a cultural response that runs counter to the values of mainstream society, one that can become potentially violent.

In May 2015, two New Orleans transplants, Ashley and Ben—middle-class young adults who are part of the New Orleans musical scene—were violently attacked in their apartment in the Seventh Ward. Although acts such as this threaten the lives of New Orleans residents, gentrification in many neighborhoods heightens levels of relative deprivation, which exacerbates feelings of discontent and injustice. Under conditions of extreme social inequality and subjectively experienced social injustice, vulnerable populations respond in various ways, some positive for the community and some not. When some members of a community—after seeing the injustices of poverty and structural exclusion—experience their neighborhood as being under siege, they may develop a violent culture of resistance. Poor and excluded residents of the urban inner city have always suffered the injustices of military threat from city officials, social marginalization and exclusion, third-world levels of poverty, and high incarceration rates. Now, to top it all off, they are confronted with wealthy white people sipping lattes and singing indie hipster music in their neighborhoods. Of course they bite back. Many develop creative cultural, religious, or political solutions. Others develop a more violent culture of resistance. As Gigi, a new resident to the Seventh Ward and friend of Ashley and Ben, recently admitted: "The dark side of New Orleans is a general presence; if you can't see it, you're blind. Crime is keeping gentrification from increasing. It's keeping us away. It scares a lot of people. If more of that shit was happening, more people would stay away."

Much of the black community in New Orleans—especially the working-class residents of some of the city's historical neighborhoods, such as the Seventh Ward, Ninth Ward, and Tremé—is perhaps down but definitely not out. From slavery to Jim Crow to racial segregation and spatial isolation to racist urban-planning policies that disrupt communities and destroy their economic and cultural vitality, the black community continues to find creative ways to lead rich and meaningful lives under grim conditions. While a minority might lead self-destructive lives that cripple their communities, the vast majority of the community residents keep the culture and vitality of the city alive. These black and Creole communities produced jazz and brass bands as well as gumbo and jambalaya, and they still do.

Many members of the community have grown increasingly distrustful of public officials who claim that the disappearance of poor black people from their neighborhoods as efforts are made to revitalize them is simply a coincidence. This is no coincidence. Gentrification is simply another form of Jim Crow. It pushes out the poor and working-class black populations from urban areas that real-estate developers and city officials see as sites for profit. The poor and working-class black and Creole community of New Orleans built the culture of the city. Now the tourist industry appropriates that culture and sells it for huge profits. The community that produced the culture is forced to leave as renters are priced out of their homes and communities. What will happen to the culture of New Orleans if gentrification continues?

Gentrification also increases the relative deprivation that many members of the black community experience as wealthy outsiders begin to move into their communities. These increases in relative deprivation hasten feelings of discontent, which can lead to building tensions and hostilities sometimes resulting in violence. We can expect street violence to continue so long as the structural conditions remain intact that create relative deprivations and feelings of discontent. Unfortunately, many of the hipster transplants, who know little or nothing about the structural conditions that create such tensions, move into historically rich communities without showing respect for or an understanding of the culture and its people. These down transplants attempt to create a hipster wonderland in poor neighborhoods, and this sometimes leads to disastrous results.

CHAPTER 9

HIPSTER WONDERLAND

The educated man pictures a horde of submen, wanting only a day's liberty to loot his house, burn his books, and set him to work minding a
machine or sweeping out a lavatory. "Anything," he thinks, "any injustice,
sooner than let that mob loose." He does not see that since there is no difference between the mass of rich and poor, there is no question of setting
the mob loose. The mob is in fact loose now, and—in the shape of rich
men—is using its power to set up enormous treadmills of boredom, such
as "smart" hotels.
—George Orwell, *Down and Out in Paris and London*

THE HIPSTERS OF NEW ORLEANS

Hipsters often travel in search of the next "authentic" bohemia to become artists, singer-songwriters, musicians, writers, poets, and other creative types,
largely refusing to work conventional jobs to support themselves. As one of
the more critical hipsters told me, "We're creating hipster wonderlands in the
ghetto." They live down and out in New Orleans, but not really. Many of these
hipsters can get out of a jam in a moment's notice with a phone call to their
parents. Unlike what I call "stylistic hipsters," who have more conventional
jobs and mainstream lives, "real" hipsters dedicate themselves to a loose set

of values that distinguish them from the mainstream world. These values include a dedication to unemployment and creative expression, largely through music and distinctive secondhand clothing. They tend to ascribe to liberal values, yet many gave me the impression of having had little exposure to people outside of their race and class. Many come from middle-class homes and possess middle-class cultural capital. They often use the term "singer-songwriter" to describe themselves.

"Music changes everything," one hipster tells me. "People have been using music to cure all kinds of shit, healing the soul and spirit. The shit's magical. It can start revolutions and change the whole fucking world, combat evil shit everywhere." Others believe their indie and urban-folk improvisations will powerfully transform the music and culture of New Orleans. Many consider themselves to be "mad" creative types teetering on the edge of sanity, sacrificing their minds and lives to gift their music to the rest of the world. Some hipster men believe, and certainly wish, that women should freely give themselves to them for the artistic gifts they provide.

Many envision forming a semi-independent and autonomous society within a society so that they can operate and produce music without having to conform to the rules and obligations of the mainstream world. Many believe in their right to receive money largely through donations. Most of the hipsters share cheap rents with other hipsters in largely poor black neighborhoods near the Vieux Carré, where they busk to make money. Many also make use of government welfare, collecting food stamps and unemployment and disability checks. By contrast, most nonhipsters on government-assistance programs such as SNAP[1] also work conventional jobs—albeit excruciatingly low-paying jobs. Those white welfare hipster queens! At least they're not taking our jobs.

Hipsters patronize coffee shops like Flora's in Faubourg Marigny, Satsuma Café in Faubourg Bywater, and Envie Cafe on Decatur Street in the Vieux Carré. They attend music venues such as Siberia on St. Claude and Circle Bar on St. Charles Avenue, where they sometimes perform in the early evenings. They find cheap eateries that once served traditional working-class communities, like Frady's Food Store and Hank's minisupermarket on St. Claude, where a dollar gets you two pieces of fried catfish. Some hipsters occasionally busk, but the more "upscale" of their ilk run speakeasy bars that act like art studios for "song shares," charge money for music events, and open pop-up restaurants set up in the backyards of their apartment buildings.[2] As a last resort, many can phone home for money. They are a loosely organized group that often depends on networking with one another to find musical gigs throughout the city, though only the same small group of hipsters ever attends any of these gigs. In almost every way, they exist in a bubble largely unaware of the city, its politics and culture, and the plight of its people. And they unknowingly serve as one of the first wave of gentrifiers before yuppies with money price them out of their cheap homes, the homes in which black people once lived prior to Hurricane Katrina. These hipsters hasten the destruction of the historically black communities of New Orleans, and they don't seem to care.

The hipsters of New Orleans are now part of the history of an old and wise city that can no longer resist the great structural transformations of our time, including the processes of gentrification and displacement associated with urban revitalization. The type of gentrification occurring in New Orleans intensifies the experiences of relative deprivation, or the subjective sense of arbitrary social injustice and unfairness in society, and this heightens feelings of discontent, producing a hostility that can lead to violent community responses.

GENTRIFICATION, RELATIVE DEPRIVATION, AND VIOLENCE

As mentioned at the end of the previous chapter, Ashley and Ben almost died during a violent break-in. But the story goes beyond the lives of Ashley and Ben and beyond the life of their accused attacker. The structural conditions creating the context that produces people willing to commit such violent acts merit consideration as well. The long historical context of urban politics and planning that is destroying the black community, the large-scale gentrification pushing black residents out of their neighborhoods, and the resistant and at times violent cultures that respond to these historical and structural forces must also be examined.

Discontent is less likely to occur when the various groups that live in the same community face similar struggles and possess similar status affiliations. The black community of New Orleans has developed a culture that creatively responds to its collectively perceived structural problems. It is a culture that shares these problems and therefore feels less discontent when inside their protective communities, which have developed for generations. The discontent many traditional black residents experience serving white tourists in the service industry or watching white yuppies drink fancy latté coffees and cocktails in and around the Vieux Carré dissipates when they return to the comforts of their traditional communities. But thanks to gentrification, such a retreat is often no longer possible. All of the yoga and lattés, the white privilege and expensive consumer choices, all the wealth and high status now comes barreling into the communities that once offered support and comfort. What is more, the moneyed and privileged roam at their own whim to settle freely anywhere they desire, and those lacking money and privilege must settle—or get violently forced—into castoff areas that the more powerful do not want. Once upon a time in New Orleans, that was the Seventh and Ninth Wards. Now, the wealthy and privileged see the Seventh and Ninth Wards as profitable city spaces ripe for the taking. And the elite urban planners will profit from the culture that was produced by those very same people who have been priced out of their community. I wonder where all the Mardi Gras Indians, brass-band musicians, and second liners will end up if this process continues. Will it be New Orleans East? Mississippi? This type of gentrification is producing a down-and-out New Orleans, and, to repeat the prevailing sentiment of our people after Hurricane Katrina, it deserves to be saved.

As many black New Orleans residents continue to experience the disrespect of increased poverty, incarceration, and exclusion within the historical context extending from Jim Crow racism to modern-day colonialism, the growing awareness of such conditions of social injustice and the general unfairness of such a society that allows this trend not only to continue but to accelerate will likely breed increased violence. This argument fits right along with Jock Young's new-left realism in cultural criminology, which argues that

> discontent is a product of *relative*, not *absolute*, deprivation. . . . Sheer poverty, for example, does not necessarily lead to a subculture of discontent; it may, just as easily, lead to quiescence and fatalism. Discontent occurs when comparisons between comparable groups are made which suggest that unnecessary injustices are occurring. . . . Exploitative cultures have existed

for generations without extinction: it is the perception of injustice—*relative deprivation*—which counts.[3]

When Ashley and Ben moved into that apartment in the Seventh Ward, they unknowingly put themselves inside a long and ongoing current of historical and structural forces. They became part of a larger history of a once-vibrant but now poor and black neighborhood that is not only still recovering from the devastation of Hurricane Katrina but also dealing with increased levels of poverty and marginalization—as well as the real threat of gentrification, which may soon displace many renters from their homes.

To get a better understanding of the simmering undercurrents that might have played a role in the attack, it is useful first to take a closer look at the small hipster community living two houses from the apartment where the attack occurred.

THE HIPSTERS OF NORTH VILLERE STREET, SEVENTH WARD

Ashley and Ben were part of an inner circle of transplant hipsters relatively new to New Orleans. Though well intentioned, these "liberal" white young hipsters rent cheap houses in New Orleans' poor and largely black neighborhoods and attempt to impart "positive" change in the community. This new hipster culture often clashes with the cultural sensitivities of a black community highly aware of the insensitivity outsiders have to the precarious economic conditions that exist in poor black neighborhoods. While these hipsters might actually believe they are benign or even beneficial forces, the reality is that many of those who move into neighborhoods like the Seventh Ward make little if any attempt to produce any substantive positive change in the community, much less start any social movement. Their youthful innocence and naiveté make them vulnerable to the harsher realities of life and can even lead to tragic consequences.

Many members of the growing hipster transplant community of New Orleans revealed to me a belief that they were creating a revolutionary social and cultural movement reminiscent of the 1950s beatniks and the 1960s countercultural hippie movement, not just aesthetically but with potential economic and political implications as well. "We can change the world," one hipster musician explains with confidence, "and our power and musical talents can surpass the generations before us and really stir up change. We are more than a movement; we are the next revolution." One of the leaders among the hipsters explains that others look up to him as a "man of his times," as this generation's Bob Dylan. Further, many suggested that their music, art, and other creative endeavors are changing the world both nationally and locally. But in reality, few people beyond their small social circles, if any, attend their public musical performances, join their informal music jams, or eat at their pop-up restaurants.

Some wonder why the local residents show little to no interest in participating in these events. For example, the restaurant pop-up Pepper Lantern was held in July in the Seventh Ward, next door to where Ashley and Ben were stabbed. The person who invented the Pepper Lantern turned the back-yard of her shared apartment into an outdoor restaurant to serve customers vegan appetizers such as a coconut and tomato gazpacho "Thai-Me-Up" and "creole" "What-A-Melon Mambo," vegan entrees such as "Shaka Taco" and "Lady Marmalade Bahn Mi," and cocktails called "ImPeachMint" and "Honey Dew Me." The music consisted of a white woman in her twenties performing flute renditions of the works of classical composers, with a short lesson about each composer between sets. Kayla, a young white hipster who was working at the Pepper Lantern, stated, "I just don't understand why some of the African-American people here don't want to come here and look at what we're doing." She was completely unaware of how others in the community might perceive the presence of these hipsters in their neighborhood, playing songs about their feelings that stem from growing up in privileged suburbia while eating "Thai-Me Up" gazpacho. Though many of these hipsters like the idea of the local community joining in their song shares (which often charge an admittance fee) and buying their arguably overpriced entrees, the fact is they socially and culturally isolate themselves from the rest of the community, with no attempt to become part of it. And that can easily be seen as disrespectful and insensitive.

RACE, CLASS, INSENSITIVITY, AND VIOLENCE

At a recent community meeting I attended on the anniversary of Hurricane Katrina, Erica Dudas, a graduate of Albion College in Kalamazoo, Michigan, and the managing director of the New Orleans Musicians Assistance Foundation (NOMAF), posed a question to a prominent local resident, Fred Johnson, a former Yellow Pocahontas Mardi Gras Indian Spy Boy and a member of some well-known black organizations, including the Black Men of Social Aid and Pleasure Club. Erica asked how black-produced local music fits "into answering some of the questions plaguing the black community in New Orleans, with specific regard to mental health and violence." In her strange statement, from a white nonnative who had moved to New Orleans in 2004 from Michigan,[4] Dudas awkwardly mixed words like "*your* community" and "*we*" when attempting to clarify the question posed to Johnson:

> I've heard that Mardi Gras Indian culture and street culture are really great for bringing people out onto the streets and getting to know *your* community, but in addition to that, *we've* got serious problems ahead of *us* in this culture. So is there, not so much "is there a plan," but what are your thoughts and reflections about how *your* culture can be used to fix some of the problems that *we're* experiencing right now?

Johnson's passionate response, quoted at length below, illustrates his reaction to a question that he perceives as ignorant at best and insensitive and disrespectful at worst:

> Let me answer that. So when you keep having the people at the top cut the bottom and cut the jobs and cut the services, that's why you have so much amplification of police killings. . . . And now you got an uprising. But, again, and I don't think you heard me: the level of *insensitivity* is amplifying, which creates a bad end result on the bottom. Because it [New Orleans] continues to have two cities, two cultures, two classes. So where's this gonna end up at? The Mardi Gras Indians can't fix that. The Mardi Gras Indians—the second-line culture can't fix that. Music is medicinal. But it's not strong enough; the medicine that's in the music isn't strong enough to fix what's economically wrong, what's socially wrong, all right? And without infusing all of this, it stems from racism. Haves and have-nots. And when you have a system of people who consistently keep their foot on the necks of other people, and you keep 'em at a low-wage-paying job, which means they always gonna live low because they make low.

In other words, after so much abuse, victims bite back. Still animated, Johnson takes further offense to the question and points out how the economic elites in New Orleans and the city planners and developers, as well as marketers from the tourist industry, exploit the culture of brass bands and Mardi Gras Indians in ways that profit only the city's elites, at the expense of the cultural bearers, those that live in places like St. Roch, Faubourg Tremé, and the Seventh Ward. "You have a small view. You can't put us all in one basket. It's like opening a box of apples, you got three bad apples, you got thirty-six apples in there, and say 'Oh, shut the box, the whole box bad.' No way. Until I get you to understand to have some level of sensitivity about how you gonna share some of that [economic prosperity from black and Creole New Orleans culture] back with this culture that you feeding off of . . ." He stops, hesitates, then states: "Everybody know how strong we are, but us." Outsiders like Erica fail to understand the structural conditions that produce social injustice and inequality; rather, such issues are ignored and tossed aside without scrutiny. The victims are blamed as the producers of such conditions and left with the expectation that they can resolve violence not through structural change but with music, which is a response to these very conditions. As Shane, a musician much admired among hipster circles in New Orleans, put it to me in a discussion about New Orleans music during a drive to the beach in Gulfport, Mississippi:

> Music is medicine. Creativity is medicine. Like the oldest fucking kind, you know. Like people were beating on drums before, hollering at the clouds before there was fucking language. New Orleans is such a direct link to that. It's like slaves were allowed to congregate in Congo Square in the 1700s and made music, African music, that has influenced everything in the United States—more or less, give or take. That's a part of lineage, that's a part of heritage is that fucking divine spark. Like the power of initiation and cosmic gods give us through music, through language, through words. Words and music, I think, it's definitely one of the most powerful forces for change. . . . You can change the world with that stuff. You really can.

That's exactly what many hipsters envision when they refuse to work conventional jobs and decide to move into poor black neighborhoods. There is no mention of dealing with the real material conditions that serve as the forces of social injustice and inequality—the same forces that allow white people to live as hipsters and start the process of excluding many hard-working but poorer black residents from renting and buying houses in their own communities. Rather, many well-meaning hipsters want to change the world through singing and playing indie music—what they consider the antidote to the structural problems of social stratification and structural inequality.

Establishing insulated "wonderlands" of hipsterdom—oases in the middle of poverty—is not only insensitive but also potentially dangerous. Without cultural knowledge and understanding, attempts at imparting "positive change" in low-income communities, establishing a "musicians' cooperative," and creating song shares and pop-up speakeasy restaurants—no matter how well intentioned—can turn dreams into nightmares of the worst kind.

HEIGHTENED INEQUALITY AND
RELATIVE DEPRIVATION ON TRIAL

The structural conditions that produce relative deprivation destroy innocent people and crush otherwise potentially innocent lives. Perhaps it's time to realize that the engines that promote gentrification breed some of the violence we see on the streets of New Orleans. We need to become sensitive to the communities of New Orleans that produce our culture. We need to become aware that though change is necessary—or at least inevitable—it need not disrespect our city, culture, and people, especially the communities that produced jazz, second lining, and Mardi Gras Indians. To end much of the violence in New Orleans, radical change must happen within the political and economic institutions that produce inequality and social injustice, derailing profit-driven elites who see black culture and geographically profitable residential areas with dollar signs in their greedy eyes. The black community in the Seventh and Ninth Wards, Faubourg Tremé, Bywater, Holy Cross, and St. Roch have the right to exist. They have the right to preserve their community and the culture that makes this city. As Jock Young says:

> If the cause of crime is injustice, then its solution must lie in this direction. If poor conditions cause crime, it must be impossible to prevent crime without changing these circumstances. Furthermore, it follows that it must be wrong to punish the offender for conditions beyond his or her control. This

would be punishing the criminal and blaming the victim. The social democratic brand of positivism, although sensing that injustice was the root cause
of crime, either deflected its attentions to purely individual deprivation (e.g.
maternal deprivation, broken homes, etc.) or made the fundamental mistake of believing that ameliorating deprivation quantitatively in an absolute sense (e.g. raising standards of education, housing, etc.) would solve
the problem of relative deprivation.[5]

If a society wants to end the type of violence that nearly killed Ashley and
Ben, perhaps the solution lies in destroying the current system of wealth and
power found in our economic and political institutions and in creating new
systems that begin by offering respect to all people. The solutions that will
lead to an end to violence rely on ameliorating the types of inequality and
social injustice that displace economically vulnerable populations from their
community and culture.

Resolving the problems associated with violence in New Orleans requires
eliminating the racially based inequality that oppresses the poor and allows
the elite continually to disregard them. As a result of this disrespect, oppressed
people, in New Orleans and around the world, will continue to resist their
own structural subordination and challenge the hostile elite class, sometimes
violently. Here in New Orleans, the oppressed will bite back so long as inequality exists and the gentrification process continues, a system and process that
exacerbate feelings associated with relative deprivation. The elites do not
suffer because of the irresponsible policies created and carried out by elite
urban planners. It's people like Ashley who suffer. It's people who live in
what was once a community of support, social safety, and comfort who are
now at the mercy of greedy landlords, raised property taxes, and exorbi-
tantly priced commodities. For those people of New Orleans who have lost
their family, neighbors, and homes—including the residents of the former
housing projects of New Orleans—I say, "Fuck Pres Kabacoff" and all those
like him.

DON'T MESS WITH GUMBO

Looking at the four levels of sociological analysis—historical circumstances,
structural conditions, cultural context, and individual biography—helps us

understand the stabbing that almost took two young lives and destroyed the life of another.

History

The historical circumstances that led to the stabbing involve the well-documented struggles associated with Jim Crow, racial residential segregation and spatial isolation, and the attempted redevelopment of a once culturally rich and economically viable black Seventh Ward neighborhood struggling with a long history of elite urban-planning policies that disrupt communities and destabilize their informal support networks.

Structure

The structural conditions involve the economic and political exclusion of a poor community of people denied access to institutional resources. This denial leaves them vulnerable to the processes of gentrification that displace long-term residents and destabilize community, family, work, and interpersonal relationships. Though economically and politically excluded, the culture of this community is further disrespected as city elites co-opt the cultural products created within communities like the Seventh Ward for their own profit, without sufficient payment or support to the culture bearers. Lacking access to institutions of power and ideology leaves community members to their own devices and creative imaginations, making it necessary for them often to take matters into their own hands to resolve the social problems that plague their neighborhoods and the lives within them.

Culture

Cultures arise as people collectively experience structural problems within the context of their everyday lives. New Orleans' black communities, such as the Seventh Ward, have historically developed some of the most creative cultural responses to their collectively experienced structural problems, including second-line parades, jazz, and brass bands. These serve as forms of both healing and resistance to structural domination. But the music is being stolen and the culture ripped from the streets of the Seventh Ward to be sold in its watered-down version in the Vieux Carré to tourists who know nothing about the city.

Those that produce New Orleans culture never intended to sell it in a tourist marketplace. The cultural gumbo that defines New Orleans embodies resilience. The tourists buying and the elites selling that cultural gumbo don't know gumbo. They can't know gumbo. But the black folks in the wards and faubourgs know gumbo—and you don't fuck with gumbo.

In the wise, experience-informed words of Fred Johnson:

> The folks who's using the word "resilient" today? They don't know nothing about resilience. They don't have an idea about resilience. They're talking about a resilience that was fueled by a lot of money that came from all sorts of places. Resilience is when you don't have no money and you know how to make something out of nothing. Take a pot of red beans and a few pickle-tips, and you put them beans together, and everybody eat as if they ate the best meal they ever ate in the world—in they life. They don't know nothing about resilience. They don't know nothing about not having toilet paper, but using a brown bag to substitute it. You don't know resilience. You don't know what it is to have to take a bath out of a bottle of water in a major city.

The perception of wealthy white people moving into poor black neighborhoods exacerbates the feelings of discontent that derive from conditions of structural inequality and social injustice. Heightened levels of discontent provide fertile grounds for harsh responses. For decades, neighborhoods like the Seventh Ward remained relatively isolated from the wealth and prosperity of the city's wealthier classes. They kept to themselves, perhaps leaving their community only to visit family, attend school, and go to work. Now there is a new discontent. It's one thing to see wealth at a distance, far removed from the personal experiences of everyday life. It's another thing to perceive outsiders flaunting their wealth in your community, the only place where many urban residents find comfort, solidarity, and support.

Lacking access to institutional resources puts residents of black communities in vulnerable positions with respect to resolving their own problems and resisting the forces of structural oppression that threaten them. Cultures develop their own unique attempts to resolve these problems. Jazz and brass bands and gumbo are some of those solutions. That's now being stolen, and the informal networks of support that once flourished in the poor black community of the Seventh Ward is being eroded by the forcing out of longtime neighborhood residents, thanks to rising housing and rental prices. As a result, gentrification undercuts the support networks and disrupts the stability of a community that people such as Darnell once depended on.

Biography

According to the established narrative, the accused Darnell stabbed Ashley in the hopes of taking her life. He looked at her and coldly said, "I'm going to kill you," and after repeatedly stabbing her belly, left the young Ashley for dead. Darnell may have made the decision to stab Ashley and Ben, but he did not make the decision in a vacuum. Rather, he stabbed her, allegedly, within specific historical circumstances and definite structural conditions where individuals operating within their cultural milieus respond with their own unique solutions to their collectively experienced problems. Perhaps with those support networks in place, Darnell would have had the resources and stability to learn better strategies to deal with the frustration and discontent that may have been at the root of the attack. Many young New Orleans kids not dissimilar to Darnell pick up trombones, not guns and knives, to deal with their problems. But the music coming out of those trombones is fading as outsiders from Anywhere, USA, move into colorful New Orleans neighborhoods. As institutions continue to deny residents access to resources and as social safety networks continue to erode, the community will feel itself under siege by urban colonialism, and discontent will reach new heights.

Residents will take it upon themselves to resolve the problems associated with these external threats and feelings of discontent. Darnell, and many other young men like him who kill and get killed in the bloody streets of New Orleans, use the only weapons left at their disposal. Sometimes, under these specific conditions, that weapon is a knife.

Without these specific historical circumstances of Jim Crow and segregation, without the structural conditions of institutional and community disruption from gentrification, and without the cultural responses that leave community members to fend for themselves to find solutions to the objective conditions of inequality and the increasing discontent deriving from it, it will be possible for people like the accused Darnell to make the decision to stick a knife into another human. Without these conditions, the act would not have been possible. Ashley would not have been cut up. Of course, Darnell should be put on trial, but the institutions of inequality and social injustice that help create the discontent that leads to violence also deserve to be held accountable.

The people of New Orleans are tied to the land and the water that runs through that land, from Lake Pontchartrain to the great curving anaconda of the Mississippi and all the bayous and waterways in between. Their blood has been spilled there; it continues to spill. Their lives have been built on that spongy loam that is the foundation of New Orleans. The power elite who run the city, the out-of-town carpetbag investors, the wonderland hipsters, and the parasitic tourist regime are not part of that swampy fabric. They don't know gumbo. They fail to understand that to exist in this fragile environment and to navigate our world, and to keep what *it* is that makes New Orleans unique, it's essential to respect the culture and the community that has been built upon her. This culture feeds the life of the city and gives her the ability to survive. That understanding and cognizance is what needs to be fostered and preserved. Like all life, it can't be abused. Until they understand gumbo, those who want to take advantage of our lifeblood need to piss off. To all you gentrifiers, as Shakespeare so diplomatically put it, "we may be better strangers."

CHAPTER 10

BRASS BANDS AND SECOND LINES

If you set yourself to it, you can live the same life, rich or poor. You can keep on with your books and your ideas. You just got to say to yourself, "I'm a free man in here"—he tapped his forehead—"and you're all right."
—George Orwell, *Down and Out in Paris and London*

IF YOU FANCY IT, you can transcend the ordinary and go beyond living as do the drab, nondancing normals. You can revel in the uncertainty of life and say to yourself, "I am here, but soon I will not be"—and think carefully before realizing it's time to dance life's second line and resist fading into silent obscurity.

And that is what New Orleans is about. We dance to save our selves, to scream "I am alive" at a late-modern world's rapid gentrification and over-developed processes of mass production that stifle and crush everything in their path into a homogeneous mass. We say, "Yeah you right" with trombones and second lines and parades, and we find creative ways to overcome our collectively experienced structural problems. Third-world poverty, Canal Street sinkholes, rapid gentrification, reified urban planning, violent and oppressive police, oil spills, wetland depletion, brain-eating amoebas, lead-contaminated drinking water, high murder rates, failing educational systems, a poverty-wage tourist industry, notoriously corrupt politicians and the thugs of City Hall, abandoned and deprived ghetto communities, and empty vacant houses, to name just a few troubles, threaten to cripple this resilient city, its people, its culture.

Perhaps the worst tragedy of modern urban life in New Orleans is the sickening expansion of poverty juxtaposed with the yuppie gentrification of the city's neighborhoods and the appropriation of its culture, which produces music, second lines, and Mardi Gras Indians. Urban-planning policies in New Orleans and the gentrification that results from it threaten this unique culture, which originated from African, Caribbean, French, and Spanish roots, among others, but which grew from the soil of Louisiana's swamps. Let's study and celebrate it at the same time. Here's a New Orleans second-line parade.

THE NEW ORLEANS SECOND LINE:
UPTOWN SWINGERS PARADE

A funeral is not a second line. All you writers, stop saying that shit. That's wrong. A funeral is not a second line. A funeral is a funeral, and a second line is a second line. You want to be so smart and so hip, and you don't know nothing. When you make that statement, you got it contorted. It's really contorted. A funeral is not a second line. Not. Do your research, do your history. You don't know, ask somebody, but stop writing that.

—Fred Johnson, director of the Neighborhood Development
Foundation, founding member of the Black Men of Labor, and former
spy boy with the Yellow Pocahontas Mardi Gras Indian Gang

I laugh and I chuckle every time I go to second line nowadays. I laugh and chuckle, because I remember back in the day, that wasn't the safest place to be. When the Indians would come out in Hunter's Field in the Seventh Ward on St. Joseph's Day, that wasn't the safest place to be. There was a lot of violence. They were serious about their culture, and a lot of machetes would come out under them Indian suits, and people would go to Charity (Hospital). But I chuckle now, I'm like, I see so many different ethnicities who are not only sitting on the sidelines of the second lines, they're

marching in the second lines. I mean, I've seen more Jewish people who
know how to buck jump. That's why I know them sons of David got a little
soul. But not only that, I'm starting to see people drumming who are not
African American. I'm now hearing that there are non–African Americans
sewing the suits for the Indians. I'm hearing that these cultures, these people are coming in, they are absorbing it and actually, I'm thankful because
this culture may just survive because the younger generations of African
Americans and the Creoles of culture don't care. And maybe this may
make them take notice.

—Vance Vaucresson, co-developer of the residential housing
project Sacred Heart Church at St. Bernard Avenue and owner
of Vaucresson's Sausage Company

Buck jumping, feet tapping and stomping, the crowd goes marching onward
and forward, and sometimes back- and sideways, toward banging drummers
and fallen skies, blood-red moons and trumpet calls, riding horsemen and
blazing fires, and the ever-marching saints into that unknown promised land.[1]
It's a moving juggernaut out of control, a rambling beautiful beast that nei-
ther gods nor man dare tame. New Orleans' living, breathing culture is not for
the lame. *Iko iko an nay*, we are on our way.

We begin the Valley of Silent Men Social Aid and Pleasure Club thirtieth-
anniversary second-line parade at Tapps II Lounge on a steamy New Orleans
Sunday afternoon in late August.[2] The brass band blows full steam on its trumpets and trombones, leading the crowds like the Pied Piper of Hamelin and
his magic pipe. But unlike Pied Piper's rats and children, these men of brass
pave the way for a determined army steeped in the Creole and black traditions of the "who dat" and "yeah you right" city. The second line's followers
do not simply follow. They become active participants of a city that practices
its culture right on its streets. Culture "way down yonder in New Orleans"
may originate from the ashes of slavery and Jim Crow, but it stubbornly persists
through a collective effort of resistance to the great oppressive structural forces
of our times.

Like a Mardi Gras Indian spy boy claiming "it cool," we smoothly buck
jump our two-way-pocky-way down Washington Avenue.[3] We head to South
Claiborne Avenue and take a left onto Jackson Avenue to Kings Fashion, in
front of which the march stops, just for a few minutes, while the collective

dance continues. A man holds a banner proudly emblazoned with the name of the Valley of the Silent Men Social and Pleasure Club ("est. 1985"), claiming that "out of the valley comes those who are chosen and they are here." The members of the club dress smartly in their fancy dress: loose-fitting white cotton suits, six-button lapelled vests, blue striped shirts with contrast collars and cuffs, and blue suede shoes with white leather soles. So attired, the men move their feet second-line style: a grooving mobile mob marching and dancing forward with umbrellas, banners, and the feathered emblem of their club. Improvising, they take turns busting buck jump fast footwork moves of stamps and chugs to a crowd watching and dancing along in delight.

The eclectic crowd largely consists of Creoles and black folks of many shades and colors, along with a sprinkle of white faces walking with the second line or dancing wildly in an effort to fit in. One tall and muscular white woman dances almost mechanically, and without pause, treating the second line as an aerobic exercise. While her moves resemble the style, it lacks the soul that only generations of history and struggle produce.[4]

We march down Jackson Avenue and eventually to Martin Luther King Jr. Boulevard, Simon Bolivar Avenue, and finally to Freret Street, heading toward the Furious Five S&P Club on the corner of Jackson Avenue. The local women of the neighborhood dress as everything from puritan Pentecostals to provocative temptresses and femmes fatales, highlighting the beautiful African features that so many New Orleans women possess. Some men dazzle in their dress; others look tough and rough with the bagging and sagging jeans and long white T-shirts that many associate with inner-city street life. Whether hot and sexy or puritanical and guarded or rough and tough or dazzled-up fancy, these New Orleans folk form a collective movement and beat that, like the Hopi rain dance, unites the diverse people into a cohesive conscious group in solidarity, even if only for a while.

Sweating and jumping, to the side, up and down, and back again, moving like tap dancers on speed while jumping almost as if to avoid bursting fireworks exploding from the ground. Busting sporadic frenzied grooves buckwilin' on the streets and in the front and back yards; some hop porches while others dance atop eighteen-wheeled trucks, waving bandanas and shaking some ass like they don't give a damn about a thing except for this exact moment. No one cares about how they look, because here in this moment it's above and beyond the private orbit of the subjective self—a rise above and beyond the ordinary to this human creative experiment we call New Orleans. We go up Freret to Philip to Magnolia to Baronne to Second Street and turn right to a place called Sportsman Corner on the corner of Dryades Street and out Second Avenue to Danneel Street, taking a left turn to Uncle P's Barbeque Stop.

As we march, images of the Mardi Gras Indian and second-line traditions pervade the memory, but a frightful thought creeps in: What if this rich culture is indeed nearing its end? Will this be the new New Orleans gentrified city that left behind its African and Creole roots and their cultural products? Will elite urban planners and developers and yuppie gentrifiers simply pay lip service to the culture while reaping profits at the expense of its producers? Will they become the new bearers of the traditions of black New Orleans culture? Surely, Vance Vaucresson is exaggerating when he describes black and Creole youths losing interest in their cultural practices. How many more years will this magic and marching and second lining and social aiding and pleasure clubbing continue? Will New Orleans become a shell of its former self, another Anywhere, USA? Our city's soul is at risk.

As the drummers beat and the trombonists blow, we move and groove and funk through the thick wet air as thoughts of gentrification, and its threat to bleach white the historically black neighborhoods, worry the mind. Who's going to produce this culture once urban postcolonial practices clear out the black and brown working and underclasses who create the raw cultural materials sold for billions in the tourist industry? When the elites provide slavery, or Jim Crow, or postcolonial gentrification, the black community of New Orleans responds with jazz and gumbo and brass bands. They also respond with the violence that continues to bloody the streets of this beloved city. Will another of our city's children be killed today?

As we move and groove up Danneel to Louisiana Avenue, stopping at Single Men S&P Club on Louisiana and Loyola Avenues, it becomes impossible not to worry that this strong and rich yet somehow fragile culture will fade, along with its practices—and eventually our memories of them. Will this history end with Eliot's whimper in some anthropologist's dusty old manuscript? What will happen to the buck jump, the gumbo, and the jambalaya? We are left wondering if these are the last remnants of a disappearing culture second lining to its own jazz funeral or if the culture will revive and awaken. This city has weathered some of the worst American atrocities, and through it all the Mardi Gras Indians and the culture have prevailed. The Indians and second liners and social clubs continue to produce a culture of gumbo resilience—a culture that now needs the collective roar of the people, all people who call New Orleans home, to save it. Will we let our soul die?

As we march and dance and chant, I wonder if outsiders and transplants will appropriate the culture and transform it into a watered-down version of cultural privilege. Or will they honor the culture and history that's endured and failed and triumphed and invented and died and continued to live in some way in and through the cultural symbols, fleurs-de-lis, impromptu funk-jazz tunes, brass bands of Jackson Square, voodoo altars and shrines, and aboveground cemeteries?

The brass-band pied pipers work their way through the streets with a long second line of hundreds of people marching forward like soulful maniacs in a culture gone beautifully mad. We march and dance down Louisiana Avenue, stopping at Diva's and Gents S&P Club on the corner of Washington Avenue and Lasalle Street and eventually continuing to Bam Bam Sister's House on South Robertson Street. As the second line finally disbands and the brass band packs its gear, it becomes clear that the future of New Orleans depends on its cultural producers and the swamp and soil in which grew the seeds of this highly unique and unusual culture, perhaps the most interesting on Earth. It rests upon the cultural health of the Mardi Gras Indians and brass bands and the young and old who keep this culture going, keep it moving and responding to the ongoing and ever-changing structural problems this swampy land faces.

THE BRASS BANDS OF NEW ORLEANS

Jazz and funk brass bands play everywhere in New Orleans. Some play on the city's streets, especially on the corner of Chartres and Frenchmen streets in Faubourg Marigny, in Jackson Square in front of St. Louis Cathedral, on Bourbon and Royal streets, near the French Market, and other areas all over the city. Other brass bands play in black-and-white events—where the servers and musicians wear tuxes—at masquerade balls and at other posh events that cater largely to the upper classes. Brass bands that acquire a good reputation often perform at music venues throughout the city, from Uptown's Tipitina's to the Central Business District's Howlin' Wolf, to Faubourg Marigny's d.b.a. and Blue Nile. Others with an established or budding reputation perform at the city's many music festivals, from French Quarter Fest to Jazz Fest. It's common for brass-band musicians to play on the streets, in music venues, black-and-white events, and festivals all in one day. Often, a brass-band musician will finish playing in Jackson Square in front of a dozen tourists to play, only minutes later, in front of a crowd of thousands at Jazz Fest.

It's a typical hot summer day in Jackson Square. The French Quarter tourist crowd stares with a peculiar trepid delight. Some even, as Washboard Brad says, relax their sphincters for perhaps the first time in their lives. Others video the performances or take pictures, their eyes wide in fascination. The brass-band players exude something much more interesting: You can clearly see the heart and soul that they put into it. But it's still just a job, they say—at least that's the practical, manifest function of the brass bands playing on the streets of New Orleans.

Tyrone's All Star Brass Band plays two sets of four songs each on Jackson Square using an assortment of brass including a bass drum with cymbal (the cymbal played with a screwdriver), tuba, snare drum, saxophone, three trombones, and five trumpets directly facing St. Louis Cathedral. Sometimes regulars meet to play in the All Star Brass Band. Other times the band consists of an assortment of other musicians waiting their turn to play, and often they just make up a name for their band on the spot. Tyrone is the snare-drum player who organizes many of the musicians for their daily performances in Jackson Square. He hustles brass music. Tyrone wears baggy jeans below his waist and a long white tank top. His girlfriend, two daughters, and mentally challenged mother-in-law often accompany him to Jackson Square. Most of the musicians in Tyrone's brass band range from their early twenties to late thirties. One of the trombone players takes a lead role while performing in front of dozens of tourists forming a semicircle around the group of talented musicians. This man leads the chants accompanying the music like "Do watcha wanna" and "You got to wind it up" while the revelers shake their bodies in delight. As the band rocks and blows, the leader introduces the musicians and walks around the crowd, dragging audience members to join the musicians. Another brass band member walks through the audience with a bucket to collect donations. They continue to attract a growing crowd of now about forty or fifty tourists, one older black man and one middle-aged white woman dancing crazily, as if possessed. Members of the crowd fill the tip buckets. After the musicians complete their first set of four songs, they take a break from the unrelenting heat. Tyrone and I talk about his life playing brass music in the streets of New Orleans.

Tyrone explains that when he was growing up, playing brass music kept him from selling dope on the streets as well as from other informal but potentially lucrative economic activities. He says playing music keeps a lot of these young men from opting to sell drugs, giving them another avenue to make money and also subvert the indignities of going to a bad school and working low-paying jobs. Although it's "just a job," it's one he loves.

> *Tyrone:* Yeah. I been coming here ever since, like, a hot minute . . . I got
> about ten years under my belt, I been coming . . . I would say about the
> last five, six years now.
> *Marina:* Yeah. You do pretty good for yourself, I can tell.
> *Tyrone:* Oh yeah, oh yeah. I love it.
> *Marina:* Do you kind of look at it just as a job or . . .
> *Tyrone:* I look at it like a job. This my job.
> *Marina:* Do you enjoy it? Honestly?
> *Tyrone:* Man, I love it. Aw, I wouldn't trade it for the world. I wouldn't give
> it up.

Tyrone is like many of the young musicians who play music on the streets of New Orleans, in that he lacks any real formal training outside of perhaps a stint in a high-school marching band. Tyrone does not read music when he performs and, in fact, does not know how to read music. He explains how the music comes from his heart and soul; it's something he feels. "I feel it. I play what I feel. When I know basics of stuff, I play it. I used to listen to a lot of CDs, like a lot of old-school, like Rebirth [Brass Band], old bands. I listen to a lot of bands, I listen to a lot of they tunes, and when I hear that we play one of they tunes or something like that, you know. I just put the rhythm in it, like, you know, put my own feel in it. That's how I do it." Although Tyrone played some trombone in a high-school marching band, he taught himself how to play beats using a pencil and an ashtray. He developed his craft until he could hang with the other musicians on the streets and eventually worked his way up to playing in the high-profile spots on Jackson Square where some of the best brass bands play. Hanging with the other cats playing brass is no easy task:

> See, I used to dance like, how you go to second lines and dance and dance. I ain't never did pick up on the drum. And when I used to be at home, I used to practice and I used to be like, "Man I want to be in a band!" So the bass drum with TBC? He the one started me, like, really started me playing music. Like, "T, we want you to play." I'm like, "Ah, y'all want me to play." So I picked up the drumming, and I play what I felt. And every day they just told me to come, and the more I played, the more I got better.

While Tyrone modestly admits that it's "just a job"—albeit one he loves—he also says it's an important source of inspiration for young men and women.

> I want to give this blessing to young brothers and show them that there's a better life in this music, you know, you can become something. Even though it's hard work coming out here every day, sweating, you know what I'm saying. But at the end of the day, it pays off. Like, your talent is what counts. And what you play from your heart, people feel that. Like they won't just look at it like, "Oh, he's playing the drums." No, they really feel that you playing from the heart, that you playing with your soul.

Besides music of the soul, Tyrone maintains that playing brass music on the streets of the city offers young men and women an opportunity not only to showcase their talents but also to become something—and maybe even make a name for themselves. This is a better alternative to the other form of equal-opportunity employment in the city—the informal economy.

> *Marina:* Do you think playing music on the streets keeps a lot of young men from doing other things, like you know, selling [drugs] and stuff like that?
> *Tyrone:* Yeah, yeah, going to jail, getting into stuff like, you know what I'm saying, getting caught up in stuff. You know, I think this music's like . . . it's like a secret.
> *Marina:* A secret.

Tyrone: A secret, you know, like, hmm. This the secret. It's real important because we got a lot of young children growing up, we got a lot of brothers that's like, young and want to be something. You know, so, I mean . . . why not play and have them come around and have they mind switched to being a musician? So it means a lot. A lot.

Marina: And you keeping the soul of the city alive while you're doing it, man.

Tyrone: Yes. Yes. I love it. Wouldn't trade it.

Brass-band music in New Orleans offers many young men in the city a chance to not only avoid the troubles and dangers of working in the informal economy but also a gig that provides the dignity and satisfaction of work—something in short supply today. What's more, it gives inner-city youths, mostly males, the opportunity to play in front of people from all over the world and even travel on occasion. Stories similar to Tyrone's happen to many other poor, black inner-city residents of the city. In a place where many black men are more familiar with the insides of jail cells than with the insides of airplanes, traveling is a big deal. As Tyrone puts it:

I had that feeling like, ahh, uhhh, ooo, sooo . . . it felt good to start playing with the big boys, you know? And then when I got that beat I just, you know, really play what I felt. And I created my own sound with them [TBC Brass Band], and made a couple of CDs and I started seeing more, like, I started traveling to places like North Carolina, started doing all kinds of stuff. . . . Man, like, it was a shock to remember, like, when we start traveling like, I never knew that I would be traveling. I never knew that I would see outside of New Orleans. I flew in the air, first time flying like, our first day we flew in the air.

Now Tyrone heads his main band, called the Twice Life Brass Band. They play in some formal events in New Orleans, like black-and-white tuxedo events at the Hilton and other gigs that cater to tourists. While we talk, a trombone player tells Tyrone about a gig in the Bywater at 8 p.m. that night, offering to text him the details. Tyrone confirms, then says, "Black and white?" to which the trombone player replies, "Nah, something different, a funeral." A man had approached this trombone player, requesting their presence at a memorial for a musician who had hanged himself.

Later that evening, about an hour or two after the memorial, some of the band members played their third gig of the day on Frenchmen Street, blowing their trumpets with the Young Fellas Brass Band for a raucous crowd dancing in the streets. Over the course of a single day, many of these brass-band musicians played in front of dozens of world travelers and tourists in front of the iconic St. Louis Cathedral, at a somber memorial in a wealthy Bywater house (right across the street from the Occult OTO Temple), and to a wild Frenchmen Street crowd of locals and tourists dancing late into the night.

OLD-SCHOOL BRASS-BAND MUSICIANS OF JACKSON SQUARE

As New Orleans continues to exclude many of the working class and underclasses in the city from educational and economic opportunities that offer living wages and personal dignity, structural strains make life harder for many inner-city residents. While the informal and illegal underground economy might offer a chance to pursue culturally desired goals, it often leads to self-destructive paths that also threaten the stability and safety of communities. The culture of New Orleans might offer poverty-wage jobs and social exclusion, but it also offers poor inner-city residents an alternative path to pursue economic opportunities and social status. It's the streets of New Orleans, especially areas in and around the French Quarter, like Jackson Square and Frenchmen Street, where poor and marginalized urban youths can make money and join in the city's musical history. A musician calling himself "Street Satchmo" and his band serve as a prime example of how these young men choose trumpets over other possibly more lucrative but destructive paths to economic and social well-being.

Street Satchmo is a large, affable man who knows how to charm a crowd, especially when pretty ladies are involved. He leads a brass band that plays soulful renditions of Louis Armstrong standards in Jackson Square to a large tourist crowd. Street Satchmo has been playing on the streets of New Orleans for more than thirty-five years. Some tourists listen in delight from the balcony of the restaurant Tableau on the corner of Chartres and St. Peter streets, while pedestrians form a semicircle around his band. Street Satchmo sings and plays the trumpet just like Armstrong, the New Orleans legend whose nickname he borrowed for himself.

> *Street Satchmo:* Years. It's been over thirty years.
> *Marina:* Yeah, I'm from here, I been seeing y'all since I was a little boy.
> *Street Satchmo:* We been out here since we were teenagers.
> *Marina:* Yeah. I was looking at nola.com, they were talking about best balcony bars or whatever, and then they said "and then they got the brass band musicians that have been out here for years."
> *Street Satchmo:* Oh yeah, yeah, all up and down, you know Frenchmen Street, too. You been there yet?
> *Marina:* Oh, yeah, yeah, yeah. I'm from here. I'm from Gentilly.
> *Street Satchmo:* Me too, Tremé 'hood.
> *Marina:* Y'all brass bands are like the heart and soul of the city, man.
> *Street Satchmo:* That's right. Everybody comes through Jackson Square. Music always been out here. It's in a hustle spot. Always a good spot to hustle in.
> *Rubin (Another band member):* We didn't want to sell drugs or do nothing else so—
> *Street Satchmo:* I'm trying to *tell* you, man.
> *Rubin:* My mama gave me a trumpet so I started playing. My family started playing out here too, so I joined the family band.

The band members, each with over thirty-five years playing music in the streets of New Orleans, reminisce about playing video games in the old arcades, cutting class, and playing in the high-school marching band—and many New Orleans high schools produce talented bands. Eventually, they started skipping school to play on the streets full time. They discovered Jackson Square as a "great hustle spot" and started hustling some of the great music of the city to French Quarter tourists. Some of the black street musicians simply followed the paths of generations before them; others used their agency to resolve the structural problems associated with poverty and low wages jobs offered in the New Orleans tourist industry. While the illicit informal economy offers access to economic opportunity to so many young people in many inner cities throughout the United States and world, the culture of New Orleans produces an alternative—and unique—opportunity for marginalized and excluded inner-city youths. Donald, another band member, says it best: "I just started playing it, making some money. Instead of selling drugs, my momma gave me a trombone." While it is naïve to think music and culture resolve all the problems of social injustice and economic exclusion, it does offer some urban youths of New Orleans a creative alternative path at least to lessen those of their personal troubles related to structural inequality. As Travis "Trumpet Black" Hill's song "Trumpets Not Guns" indicates, many inner-city youths of New Orleans use their agency and pick up trumpets instead of choosing more violent paths.[5]

CORPORATE THEFT OF SOUL CULTURE

Many people hold on to the notion that the urban poor are the main perpetrators of crime. These people believe that the underclass of the city steal and rob others to get easy money for personal gain. They couldn't be more wrong. Contrary to popular opinion, the rich are the ones who steal from the poor for personal gain and greed. This is no more evident than in New Orleans, where the culture of the poor is appropriated from the historically black communities only to be sold by the rich and powerful in the French Quarter's tourist industry. Consider New Orleans music from jazz to brass bands or its cuisine from gumbo and crawfish etouffée to shrimp creole and po-boys, or its cultural practices from second lines and jazz funerals to the Mardi Gras Indians. Fred Johnson, the director of the Neighborhood Development Foundation in New Orleans, founding member of the Black Men of Labor, and former spy boy with the Yellow Pocahontas Mardi Gras Indian Gang, points to the fact that young black kids grow up making the music that is sold nationally to promote New Orleans culture. He says, "The boys never see a dime of that profit. They shit on the musicians, and they shit on the culture." All of these cultural artifacts are produced in the historically black neighborhoods in the city but sold in a watered-down version to tourists in the French Quarter. While the city's largely poor and working-class residents produce the culture, the city's elite tourist regime co-opts these cultural products and sells them for its own profit. And as they get wealthier, the cultural producers of the city become poorer, and inequality continues to increase at alarming rates. And as city residents become aware of the relative deprivation that results from the elites profiting from a culture they did not produce, feelings of discontent heighten and boil over, often resulting in violence and destruction. There is no indication that the city's elites plan to resolve this contradiction between cultural producers and those that profit from it or that the city plans to find any real solutions to the structural problems of poverty and growing inequality. In fact, if there is a solution, it seems that it is to remove the working classes and underclasses through prison and gentrification. The culture of New Orleans depends, in large part, on what happens to the historically black communities of the city that produced and continue to produce the Mardi Gras Indians, brass bands, and all the gumbo. Kill that with gentrification and stifling poverty and inequality, and the city dies with them.

THE FOGS OF NEW ORLEANS AND THE FUTURE OF THE CRESCENT CITY

It is a feeling of relief, almost of pleasure, at knowing yourself at last genuinely down and out. You have talked so often of going to the dogs—and well, here are the dogs, and you have reached them, and you can stand it. It takes off a lot of anxiety.

—George Orwell, *Down and Out in Paris and London*

IT IS A FEELING OF LIBERATION AND OF TERROR, realizing that not only do your worst fears exist but that they don't matter and you are irrelevant to it all. Such thoughts have always seeped into your consciousness, but only for one horrific moment at a time, until the illusions of reality quickly restore and provide security. Here you are, drowning in existential anomie, or residing alongside Orwell's metaphorical dogs, and you realize you *can't* stand it. It fractures your reality. Now you begin finally to live.

A thick Caribbean fog from the Gulf of Mexico descends upon the French Quarter, illuminating and obscuring the magic and mystery of the enchanting Crescent City. The fog from the city blurs images, as in an old painting. The images of the French Quarter's historic buildings appear and disappear through the fluctuating lights. Living forms move and flow amid the layers of the murky cloud as they traverse the interstitial and highly fluid spaces where urban life transpires. An oppressive feeling weighs upon the body and soul—damp and cold and heavy—while walking through the misty thickness. But this fog offers a more accurate picture of the city than the images on picture-perfect postcards. This fog at once conceals like a cloak, appears as ominous as the calm before a hurricane, and exposes the naked vulnerabilities we try so hard to keep buried within.

This walk through the darkness produces an odd disconnect; one becomes unstuck in time and space, like a transcendental body adrift, floating with other celestial beings through the murky dense whiteness. The city appears somehow more real in this surreal state. The fiery hues of the moon combine with the light of the antique gas lamps that pierce the dense fog of the French Quarter.

The lights cutting through the fog cast shadows that simultaneously highlight and conceal the city's hidden secrets, but only in glimpses and moments. The eyes capture mere specks of time, disguising and revealing truths, exposing images, blurring figures, obscuring faces and movements, leaving only fleeting shadows in the night. The light emerges from the permeable fog, blurring the hard and soft city, hiding its details, muddling the distinctions between its people, spaces, objects, and architecture. Every corner offers a new reality, a vacuum through time and space, as solidity and liquidation become one and the same.

The sounds of the city fade into the background; its smells take precedence in stimulating the senses. At least I'm still alive right now—that is all there is to be sure of at this moment. Those who get to know the city this way become alive for the very first time. They do not merely see it; rather, New Orleans must be experienced through all the senses. Here in New Orleans you hear the culture, see its aroma, smell its colors, taste its music, feel its architecture, and touch its spirit. Human senses betray and take on new unexpected dimensions, which cross over and blend into one another. You become New Orleans, and it becomes you. You become the city's fog.

The ambiguity and uncertainty of the shadowy fog of urban life hastens feelings of anomie while also providing a feeling of existential relief, an unhappy pleasure, that the world is nothing like it has presented itself and that reality has always been humanity's secret illusion. The safety blanket of comforting reality that keeps truth hidden at bay fades as the steamy fog of the New Orleans night at once reveals the illusions of reality. Only then do you discover life. It exhilarates and terrifies the deepest part of the soul. You commit the opposite of suicide, a free fall in reverse.

As the fog masks and highlights, darkens and softens, purifies and taints, sanitizes and muddies, clarifies and obscures, we see through it images and shades of characters appearing and disappearing only to appear again. We see Tim the Gold Man searching for relief at the end of a beer and somehow making a living in the informal streets of a city that provides him with semi-celebrity status. In the spaces between the fog we see people like Cubs the Poet writing for tourists, Eric Odditorium swallowing swords, and Stumps the Clown swinging a bowling ball through hooks in his ears, all attempting to stake their claim in the world, assert their identity through trades and skills that allow them to find meaning on the urban fringes of the city. The light filtering through the fog illuminates the three dimensions of Shannon's life as she moves from peacefully writing poetry about the French word for orgasm to whipping tourists frantically on the streets to hopping freight trains in a quest to reach the summit of life's experiences. The fog thickens to hide the occultists and Satanists as they enter the swamps of Louisiana and the cemeteries of the city to engage in supernatural rituals that empower the self and give meaning to their world. The light pushing through the ebb and flow of the New Orleans fog lights up the vibrant scenes of buskers performing on the streets and then fades out as others move through the darkness into one of the city's many abandoned houses to squat for the night.

Between and through the mist we capture the brass-band musicians blowing their trumpets and trombones, making money from tourists while also keeping the culture of the city defiantly alive. The ever-present Mardi Gras Indians of the past, present, and future marching strong in their second line and singing, "Hey now, hey now, Iko iko, iko iko unday. Jockomo feeno, ah na na, Jockomo feena nay."

We see other brass bands screaming "Fuck the po-po" with their middle fingers raised, speaking truth to power about police brutality. An opening in the fog directs our eyes to the Saturday-night lights of the Hi-Ho Lounge, where no more locals are to be found. A gaping hole in the fog reveals for a moment a money-grubbing developer concocting schemes to gentrify the next neighborhood, sucking the soul of New Orleans culture and clearing out the neighborhood locals who produce it. A few blocks away, we witness longtime residents and business owners of the community, like Vance

Vaucresson, getting the okey dokey from urban planners and city elites hell-bent on transforming the culture of New Orleans into personal profit. And there goes the defiant Fred Johnson, teaching the privileged that resilience has nothing to do with rich people maximizing their institutional resources—rather, it's about poor people finding creative ways to overcome the most oppressive of conditions. The fog casts shadows that reveal and obscure the actions of heroes and monsters that shape the city for love of the culture or greed and financial gain. One way or another, monsters are real in New Orleans.

The fog gently dissipates, exposing the violence of the city and Darnell's bloody blade repeatedly stabbing Ashley and Ben. The density suddenly thickens to hide the structural conditions that explain such acts of violence, including the processes of gentrification in historically black neighborhoods, processes that disrupt communities and hasten feelings of discontent associated with relative deprivation in an era of increasing economic inequality. The fog opens again just for a moment to show us Darnell's dead body swinging on the end of a rope while in confinement in East Louisiana State Hospital, nearly eleven months after he allegedly spilled the blood of others.

But monsters come in many forms—from urban killers to corrupt cops to economic and political city elites hastening gentrification in an era of urban postcolonialism. The fog clears momentarily as the brass bands play to revelers celebrating this highly unusual and somehow intact culture—and then it quickly closes in again to hide the carpetbaggers who threaten to destroy it all. Peer long enough and hard enough into the fog, and you'll see the public voodoo temples of outsiders such as the wife of the millionaire Pres Kabacoff—the "Voodoo Queen" Ann Glassmen—and the so-called healing center that appropriates New Orleans culture as it "makes groceries" and "makes communities."

Monsters have always been real in New Orleans. The distant images of the slave torturer Delphine LaLaurie and the jazz-loving Axman killing the innocent fade as spaces in the mist reveal the city's new monsters: Sydney Torres personally funding armed agents of social control to clear the city streets of the homeless; Pres Kabacoff pursuing profit and greed, bulldozing the urban black neighborhoods.

The fogs of eternity link time and space, connecting New Orleans to the fogs of Paris where Orwell lived and breathed among the most downtrodden.

Orwell found pleasure in becoming down and out, learning that freedom exists in one's head; he also discovered that poverty and employment produces blank, resourceless minds. He found a condescending culture of privileged people who can preach to the poor simply because the latter are perceived as having less value in a capitalistic system. The rich and privileged actually believe they deserve their wealth and status. He observes that, aside from their incomes, there is no inherent difference between urban beggars and ordinary "working" men, or between the rich and the poor. Orwell discovered a world that hated the poor and that viewed the down and out with a particular repugnance—yet also with fear of their revolting against their deplorable and inhumane existence. He knew that the rich remained privileged only because of the existence of the poor. He also realized that so-called intelligent and cultivated people with liberal opinions, like many university professors today, spend little, if any, time with the poor. Regardless of their studies and liberal attitudes, educated people know almost nothing about poor people, except what they read in books. Yet, as Orwell observed, the "average millionaire is only the average dishwasher dressed in a new suit." Perhaps deep down inside they know it. Maybe we all know it. Orwell also pointed out that money is the main test of human virtue in our society, as sick as it is, and it makes monsters out of those who command it. The monsters are now bigger than ever, and they have sharp teeth.

Though separated by an ocean and a century, we find striking similarities between the urban underground of 1920s Paris and New Orleans' postmodern urban scene. Religious men in homeless shelters still preach to the poor, and self-identifying vigilantes pay goons to rid the streets of them. We see city elites simultaneously pay lip service to black and Creole culture while excluding their access to political and economic power and resources. We also find a world of the urban poor beyond the descriptions found in the quantitative manuscripts of criminologists, which paint a picture of the poor as mundane and pathetic. Instead we find people who yearn for love and fulfillment, find creative ways to solve problems, fall in love, chase dreams, invent music, look to supernatural entities for answers, and feel the same pulsating and beating life of romance and joy and mystery and fear—perhaps even more so. For they are freed from conventional thoughts that trap us in bubbles that we believe are real, no matter how false we know them to be. They also bloody their streets as they struggle to survive under the oppressive conditions of American apartheid.

The clouds of fog open for a moment, like the biblical parting of the Red Sea, providing a path for my annual pilgrimage to my childhood home on 2342 Odin Street, in Gentilly. Today it's little more than a plot of land with unkempt grass, but the magical New Orleans fog that connects time and space produces images of my father putting on his bulletproof vest late at night to head to work in his undercover 5.0 Mustang. How I hope Pop comes back alive. There's my sister in the small, fenced-in backyard, as a child, planting that tree. It still stands today, the only shred of life, besides the grass, on the property that Hurricane Katrina spared. I see my brother cutting the grass along the side yard, and there go our dogs Peaches and Kelly playing in the backyard without a care in the world. There's me and my father, desperately clearing the gutter to prevent the rising water from entering our home, and there's my mother cooking Gagi's chicken soup, making the world seem safe again, if only for a moment.

Now the water pours in.

We are all drowning.

There are the waters of Hurricane Katrina rising above our Odin Street home, tearing away my childhood and robbing me of all but my memories. Suddenly my soul leaves my body to watch from above as my physical being clutches my hands to the soils of Odin and Spain streets, the soils of my home, the soils of my New Orleans, as I try desperately to hold on just a little longer—until it leaves my clutching grasp as the fog takes me to a new time and place. There go my mother and father, boarding the plane from Cuba as children and landing in New Orleans as exiles in a foreign land. I realize how my family lost its culture as we were torn away from our land and people—from Cuba and its stubbornly resilient culture, which stood up against imperialism, no matter how disastrous the results. We lost our souls in Cuba but found them again in the culture of New Orleans. I now realize that New Orleans saved me, too. It's more than home—it's the fabric of my being.

New Orleans, you beautiful beast, you saved so many of us who call you home, even though you can be so unapologetically brutal.

The fog lifts and produces a New Orleans rain that falls hard and heavy, but only for a moment. Perhaps the rain offers salvation to all the harm we have done to our city—from the clearing out of our wetlands that left us vulnerable, to our treatment of the communities that produce the heart and soul of the city that we now all share, for those of us who can feel its pulse.

New Orleans does not exist entirely in its own fog. It's an American city with many of the same urban problems that exist in a globally connected world. Like other cities, New Orleans is experiencing an accelerated gentrification process, sharp increases in poverty and inequality, shocking street violence, increased racial segregation, rapid immigration, growing unemployment, and heightened distrust in public officials. Although New Orleans is a unique city, many cities today share its central problems. As the fog clears, it reveals that these very same problems threaten any number of metropolises in the United States and beyond. Our cities are becoming playgrounds for the rich and powerful and for the upper-middle-class yuppies who can afford the rising rents and high costs of living—but they, too, will be priced out of the cities. We as a society, as a people, have lost control of one of our greatest human experiments and arenas for self-expression and creativity. The wonder of the mighty metropolis is being diminished and commodified. In our postcolonial era, political and economic elites now economically and politically colonize our cities, taking control of our land, resources, and institutions.

Our cities have become increasingly reified—no longer a product of our human subjectivities—and made in the image of the powerful urban regimes taking control of everything from our schools and public parks to our culture and urban spaces. This process will continue until we end their power and control, take back our cities from the greedy hands of the hyperwealthy—and those who support them—who lack any concern for the well-being of people and communities. The city belongs to the people who have created it from the ground up, who bled in it, and who still actively produce the culture from within their communities. This is a call for the people of all cities to wrest control from the superwealthy and save our threatened urban environment, including its spaces, its culture, its institutions, and its very soul. The problems in New Orleans threaten you, too, urban America.

While the rest of the urban world shares many of the same problems as New Orleans, the extraordinary Crescent City remains a unique place steeped in the cultural influences of Africa, the Caribbean, and Europe, mixing and blurring and combining in ways made particularly distinctive as it grew from the swampy soil of southern Louisiana. It's the place where Mardi Gras Indians still chant defiantly in their hand-sewn suits; where brass bands still blow their horns, trumpets, and trombones on the city streets; where citizens hold jazz funerals and second line in the streets to celebrate life and death; where red beans and rice are consumed on Mondays and just about everybody loves crawfish boils; where people work their jobs not for career advancement but to fund all the food and festivals New Orleans offers. No one cares who you are in New Orleans, and you can become whatever you want—so long as you respect the culture.

New Orleans is also a place where life on the urban fringes, deep in the cultural underbelly of the city, can not only still exist but also flourish. As wealth continues to trickle to the top in this inverted state-sponsored capitalistic system, the working and middle classes continue to tumble down the social and economic ladders. Perhaps we will see fewer willful urban outsiders navigating these urban fringes and more of the middle and working classes fall from economic grace into this underbelly. As the world's people and resources have been exploited and colonized, our political and economic leaders have turned inward to exploit their own people and colonize their homelands. We are the new and final group to be colonized and exploited. We are quickly becoming the new third world as the political and economic elite continues to distance themselves from us.

On the other hand, perhaps this flourishing cultural underbelly of everything from brass bands to buskers will lose control of the cracks and crevices of the city spaces as the political and economic leaders take over the streets and appropriate the culture only to sell it all away.

Now is our time to save New Orleans. It's time to save it from the new monsters of modernity that wreak havoc on our land and culture. Only the people of the city can prevent New Orleans from becoming down and out.

But to save New Orleans, we must know her better, become sensitive to her fragility and the precariousness of the culture that gave us our soul. We must listen to the beat of her drums and sousaphones, listen to her old and wise citizens, touch the soil, and impart our subjective imprint onto the city as we externalize our very being, our essence into the living and suffering culture currently in peril. It's life in the cultural underbelly of New Orleans— the brass bands and Mardi Gras Indians, the street performers and buskers, the voodoo practitioners and occultists, the gutter punks and urban campers, the social aid and pleasure clubs and second liners, the jazz and zydeco musicians, the Frenchmen Street poets and food trucks, sword swallowers and street mimes, the willful urban outsiders and transgressive misfits, and even all the bohemians and hipsters who maintain and continue to produce this culture. The down-and-out people of the cultural underbelly of the city have always kepxt the soul of the culture alive. Ironically, it may very well be the new down-and-out urban dweller city transplants that carry the culture into the future and save the city.

As the fog lifts, it becomes clear that now it's up to all of us who call New Orleans home to help keep this culture alive—or risk losing it all.

NOTES

1. NEW ORLEANS: ROMANCING THE CITY OF SIN AND RESISTANCE

1. A New Orleans shotgun is a style of house so named because one can open the front door and shoot a gun straight through to the back door, with the shot passing through each room on the way. In New York City, they're called railroad apartments. The design may have originated in Haiti.
2. Unless otherwise specified, the characters in this book were given aliases, or, at the request of the research participant, only their first names were used if they were common names that did not reveal their real identities.
3. The Quarter may have been one of the first urban neighborhoods in the United States to gentrify (back in the 1910s and the 1920s). See Arnold Hirsch, "New Orleans: Sunbelt in the Swamp," in *Sunbelt Cities: Politics and Growth Since World War II*, ed. R. M. Bernard and B. R. Rice (Austin: University of Texas Press, 1983), 100–137.
4. New Orleans, from its very beginning, possessed a rare agency and resiliency that eventually formed it into the fascinating sociocultural experiment that it is today. For an excellent historical account of the improvisation of this great city, see Lawrence N. Powell, *The Accidental City: Improvising New Orleans* (Cambridge, Mass.: Harvard University Press, 2012).
5. Walter Benjamin, *The Arcades Project*, trans. Howard Eiland and Kevin McLaughlin (Cambridge, Mass.: Belknap, 2002).
6. So did Henry Miller, but during the Depression of the 1930s.
7. Carlos Baker, *Hemingway: The Writer as Artist* (Princeton, N.J.: Princeton University Press, 1972).

8. Orwell uses the French word "*plongeur*" in *Down and Out in Paris and London* to describe the low "slave of slaves" jobs in the restaurant industry that include, among other things, washing dishes and serving the restaurant employees.

9. See John Shelton Reed, *Dixie Bohemia: A French Quarter Circle in the 1920s*, Walter Lynwood Fleming Lectures in Southern History (Baton Rouge: Louisiana State University Press, 2012).

10. John Barry, *Rising Tide: The Great Mississippi Flood of 1927 and How It Changed America* (New York: Simon & Schuster, 1998).

11. Angela Carll, *Where Writers Wrote in New Orleans* (Donaldsonville, La.: Margaret Media, 2013).

12. In this quotation, Sherwood Anderson, the novelist and author of *Dark Laughter*, describes 1920s New Orleans. See Reed, *Dixie Bohemia*.

13. New Orleans celebrates the annual Tennessee Williams Literary Festival, which includes a contest recognizing a scene from *A Streetcar Named Desire*, in which participants take on the role of the play's lead, screaming "Stella" from the street below to a woman in a balcony.

14. The sazerac is a classic New Orleans whiskey-based drink with Peychaud bitters, sugar, Herbsaint (or absinthe), and lemon peel.

15. See Sherwood Anderson, *Dark Laughter* (1925; Mattituck, N.Y.: Amereon Limited, 1983); Carll, *Where Writers Wrote in New Orleans*.

16. Jack Kerouac, *On the Road* (1957; New York: Viking, 1997).

17. Matt Sakakeeny, "Why Dey Had to Kill Him? The Life and Death of Shotgun Joe Williams," *Louisiana Music* (Winter 2012): 143–48.

18. Jonathan Raban, *Soft City* (New York: E. P. Dutton, 1974).

19. "Banquette" is New Orleans vernacular for sidewalk. A "neutral ground" is what other Americans refer to as a median or the grassy area between the paved lanes of a street, avenue, or boulevard. Canal Street was the city's original neutral ground, and in much of the colonial era of the eighteenth century served as the dividing line between the city's American and Creole populations, who were at odds. "Lagniappe" is a common New Orleans word that originates in French Creole meaning a little something extra on the side for free. Locals generally expect a little something extra when they frequent their favorite restaurants and bars. Some New Orleanians name their dog Lagniappe.

20. Sometime after this writing, Mike Scott of New Orleans' *Times Picayune* newspaper pointed me to a *New York Times* article on New Orleans exiles living in New York stating, "Like Parisians, many children of New Orleans regard residency in any city but the one they were born in as a term of exile." Penelope Greem, "How NOLA Can You Go?" *New York Times*, February 25, 2015, http://www.nytimes.com/2015/02/26/garden/a-new-orleans-couple-decorates-in-brooklyn.html.

21. See Ned Sublette, *The Year Before the Flood: A Story of New Orleans* (Chicago: Lawrence Hill, 2009).

22. "The City That Care Forgot" is a sobriquet for New Orleans that is believed to have originated in *Visitors' Guide to New Orleans* (New Orleans: Waldo, ca. 1879).

23. Although "Vodou" is the proper way to spell the word to keep consistent with the religious practices of Haiti, New Orleanians typically use the spelling "Voodoo."

24. "Nomic collective isolation" is my concept, referring to the momentary experiences of simultaneously being part of a strong group's collective consciousness while also being completely alone "inside" one's self.

25. Richard Campanella, "Guest Editorial: What Does It Mean to Be a New Orleanian?" *Times Picayune*, December 8, 2013.

26. Some New Orleans residents believe this to be the place where a young man named Louis Armstrong was given his first horn after being arrested in 1913. Armstrong attended the Colored Waifs' Home, later called the Municipal Boys Home, which eventually merged with Milne Boys' Home on Franklin Avenue in the Gentilly neighborhood.

27. In New Orleans, dressed po-boys include lettuce, tomato, and mayonnaise (pronounced "my-nez").

28. Karl Marx, *Writings of the Young Marx on Philosophy and Society*, ed. and trans. Lloyd David Easton and Kurt H. Guddat (Garden City, N.Y.: Anchor, 1967).

29. See C. Wright Mills, *The Sociological Imagination* (New York: Oxford University Press, 1959); and the notorious scoundrel to mainstream criminology: Jock Young, *The Criminological Imagination* (Cambridge: Polity, 2011).

30. David C. Brotherton and Luis Barrios, *Banished to the Homeland: Dominican Deportees and Their Stories of Exile* (New York: Columbia University Press, 2011).

2. THE HARD AND SOFT CITY: A PORTRAIT OF NEW ORLEANS NEIGHBORHOODS AND THEIR CHARACTERS

1. See Jack Kerouac, *On the Road (The Original Scroll)* (1959; New York: Penguin, 2008). In Kerouac's *On the Road*, the concept of *fellaheen* people is used to describe the indigenous or disenfranchised and primitive peasant class of people tied to the land, mystically aware of existence but somehow innocent to the corruption of civilization. Kerouac used this term when arriving at the end of the road in Mexico.

2. I develop the term "normals" to refer to those people who have internalized the ideas and values of mainstream society and its taken-for-granted assumptions, along with its many ideologies that uphold power and domination. Indeed, many "normals" develop a sort of Stockholm syndrome, often uncritically adopting the beliefs of their political and economic masters. Some even become *kapos*, enforcing the will of their rulers on their own brother and sister proletariat.

3. Jonathan Raban explains his concept of the soft city: "The city as we imagine it, the soft city of illusion, myth, aspiration, nightmare, is as real, maybe more real, than the hard city one can locate on maps in statistics, in monographs on urban sociology and demography and architecture." Jonathan Raban, *Soft City* (New York: E. P. Dutton, 1974).

4. Jock Young once called me delightfully insane, a great compliment coming from such a rare and peculiar and beautiful creature who, even after his passing, makes the mainstream criminologist, at times, tremble. Thanks, Jock.

5. Georg Simmel, *The Sociology of Georg Simmel*, trans. Kurt H. Wolff (N.p.: Nabu Public Domain Reprints, 2012).

6. "Crackerjacked" is my term that refers to the inflated prices middle- to high-end restaurants charge. The idea is that middle- and upper-class people seem to believe that there exists a correlation between food price and quality, the higher the cost for an entrée the better the quality of food. In one restaurant on Decatur Street, a plate of chicken was cleverly dubbed "yard bird," with a corresponding price of $25. Crackerjacking may also serve as a multicultural neoliberal tactic: frequenters of multicultural restaurants can consume authentic cultural cuisines without having to deal with an uncomfortable minority and low-income social atmosphere.

7. Since Hurricane Katrina, New Orleans has experienced a steady increase in visitors from tourism and conventions as well as increases in visitor spending. According to the New Orleans Chamber of Commerce, New Orleans set a new record in visitor spending in 2015, hosting almost ten million tourists, who generated slightly over seven billion dollars in revenues. New Orleans culture—from its brass-band music and cultural festivals, to its spicy foods and mixed cocktails, to the peculiar behaviors of second lining and jazz funerals—attracts the tourist. New Orleans Chamber of Commerce, "New Orleans Sets New Record for Visitor Spending in 2015" news release, May 5, 2016. According to the New Orleans Chamber of Commerce, this 2015 New Orleans Area Visitor Profile study was completed by the University of New Orleans (UNO) Hospitality Research Center for the New Orleans Convention and Visitors Bureau (CVB) and New Orleans Tourism Marketing Corporation (NOTMC). http://www.neworleanschamber.org/news/details/new-orleans-sets-new-record-forvisitor-spending-in-2015. The seven billion dollars in visitor revenues surpassed the 2014 mark of just below seven billion. According to the New Orleans Convention and Visitors Bureau, "In 2014 New Orleans welcomed 9.52 million visitors who spent more than $6.81 billion, the highest visitor spending in the city's history." New Orleans Convention and Visitors Bureau, "What's New in New Orleans' Hospitality Industry: February 2012," http://www.neworleanscvb.com/press-media/press-kit/whats-new/.

8. New Orleans Convention and Visitors Bureau, "New Orleans Named One of the Best Big Cities in America by *Condé Nast Traveler*," October 22, 2015, http://www.neworleanscvb.com/articles/index.cfm?action=view&articleID=9557&menuID=1604.

9. See David L. Gladstone, "Event-Based Urbanization and the New Orleans Tourist Regime: A Conceptual Framework for Understanding Structural Change in U.S. Tourist Cities," *Journal of Policy Research in Tourism, Leisure, and Events* 4, no. 3 (2012): 221–48; David Gladstone and J. Préau, "Gentrification in Tourist Cities: Evidence from New Orleans Before and After Hurricane Katrina," *Housing Policy Debate* 19, no. 1 (2008): 137–75; A. Hartnell, "Katrina Tourism and a Tale of Two

Cities: Visualizing Race and Class in New Orleans," *American Quarterly* 61, no. 3 (September 2009): 723–47.

10. Allison Plyer, Nihal Shrinath, and Vicki Mack, "The New Orleans Index at Ten: Measuring Greater New Orleans' Progress Toward Prosperity," The Data Center, July 31, 2015, http://www.datacenterresearch.org/reports_analysis/new-orleans-index-at-ten/.

11. Vicki Mack, "New Orleans Kids, Working Parents, and Poverty," The Data Center, February 26, 2015, http://www.datacenterresearch.org/reports_analysis/new-orleans-kids-working-parents-and-poverty/.

12. See U.S. Census Bureau, "Quick Facts: Orleans Parish, Louisiana," http://www.census.gov/quickfacts/table/PST045215/22071. See also Carmen DeNavas-Walt and Bernadette D. Proctor, "Income and Poverty in the United States: 2014 Current Population Reports," United States Census Bureau, September 2015.

13. The old-money elites of New Orleans frown on both the poor and scoff at the upper middle class and new upper classes with their "new money" earned from gainful employment.

14. "America is the wealthiest nation on Earth, but its people are mainly poor, and poor Americans are urged to hate themselves. . . . It is in fact a crime for an American to be poor, even though America is a nation of poor. Every other nation has folk traditions of men who were poor but extremely wise and virtuous, and therefore more estimable than anyone with power and gold. No such tales are told by American poor. They mock themselves and glorify their betters." Kurt Vonnegut, *Slaughterhouse-five: Or, The Children's Crusade, a Duty-Dance with Death* (New York: Delacorte, 1969).

15. Pierre Bourdieu, *Distinction: A Social Critique of the Judgment of Taste*, trans. Richard Nice (Cambridge, Mass.: Harvard University Press, 1984).

16. See Gladstone, "Event-Based Urbanization"; David Gladstone and J. Préau, "Hard Times in the Big Easy: Disaster Recovery in a Tourism-Dependent City," unpublished paper presented at the fifty-fifth annual Association of Collegiate Schools of Planning (ACSP) in Houston, Tex., 2015.

17. See Jane Jacobs, *The Death and Life of Great American Cities* (1961; New York: Modern Library, 1993).

18. Mitchell Duneier and Ovie Carter, *Sidewalk* (New York: Farrar, Straus and Giroux, 1999).

19. Raban, *Soft City*.

20. Few New Orleanians agree on the exact boundaries of neighborhoods. Statistical data banks also draw up boundaries within and between neighborhoods differently. As a result, while some of the demographics of the neighborhoods provided below give a more complete profile—like that of the French Quarter, Marigny, Bywater, and Tremé—the descriptions of much larger neighborhoods like Uptown and Mid-City provide a sample of the area's demographics.

21. A New Orleans native I know calls strip clubs "Sallie Mae establishments," referring to the many young women who work such jobs to pay back student loans.

22. The data for this paragraph was obtained from http://www.datacenterresearch .org/data-resources/neighborhood-data/district-1/french-quarter/.

23. The data for this paragraph was obtained from http://www.datacenterresearch .org/data-resources/neighborhood-data/district-7/Marigny/.

24. The data for this paragraph was obtained from http://www.datacenterresearch .org/data-resources/neighborhood-data/district-7/Bywater/.

25. Frady's One Stop Food Store sells po-boys and lunches. It is painted a deep banana yellow and decorated with signs, some of which are "Eat at Joe's," "Eat with King's," "Plate Lunch," "Poor Boys for all kinds," "Shut up and EAT!" with a painting of Elvis by Dr. Bob. It sits at 3229 Burgundy at the corner of Piety. It was a fresh market in 1890, turned to a deli in 1960s, and was bought and transformed in the 1970s into its present incarnation.

26. For a fascinating historical overview of the neighborhood's history, see the documentary film *Faubourg Treme: The Untold Story of Black New Orleans*, http://www .tremedoc.com.

27. The data for this paragraph was obtained from http://www.datacenterresearch .org/data-resources/neighborhood-data/district-4/Treme-Lafitte/.

28. Herbert Asbury, "Gambling on the Western Rivers," in *Sucker's Progress: An Informal History of Gambling in America from the Colonies to Canfield* (1938; Fort Wayne Public Library, 1956); Herbert Asbury, *The French Quarter: An Informal History of the New Orleans Underworld* (1936; New York: Basic Books, 2003); Al Rose, *Storyville, New Orleans: Being an Authentic, Illustrated Account of the Notorious Red-Light District* (Tuscaloosa: University of Alabama Press, 1978).

29. As described, and quite accurately according to various historical accounts, in Barbara Hambly's mystery book *Dead Water* (Ealing: Bantam, 2004).

30. The data for this paragraph was obtained from http://www.datacenterresearch .org/data-resources/neighborhood-data/district-1/central-business-district/.

31. The data for this paragraph was obtained from http://www.datacenterresearch .org/data-resources/neighborhood-data/district-4/Mid-City/.

32. Much of this section derives from information obtained in the greater New Orleans Community Data Center's Lower Ninth Ward Neighborhood Snapshot, http://www.datacenterresearch.org/pre-katrina/orleans/8/22/snapshot.html.

33. The data for this paragraph was obtained from http://www.datacenterresearch .org/data-resources/neighborhood-data/district-8/Lower-Ninth-Ward/ and http:// www.datacenterresearch.org/data-resources/neighborhood-data/district-8/Holy-Cross/.

34. As the Tulane geographer Richard Campanella says, "Things at odd angles tell interesting stories. The New Orleans cityscape abounds in such eccentricities— misaligned streets, odd-shaped blocks, off-axis houses—and like archeological artifacts, they shed light on decisions from centuries ago. Such is the case for one of the most peculiar quirks of our map, a dizzying labyrinth of streets in the heart of the historic 7th Ward." "Plantations, a Pepper Sauce and the Peculiar History of the

7th Ward 'Labyrinth,'" http://www.nola.com/homegarden/index.ssf/2015/04/plan tationscanalsapeppersa.html.

35. Much of this data is obtained from http://www.datacenterresearch.org/pre-ka trina/orleans/4/14/snapshot.html.

36. A local community resident named Armand Claiborne, quoted in the *Times-Picayune*, recalls that Claiborne Avenue was once called the black people's Canal Street. "Planners Push to Tear out Elevated I-10 Over Claiborne," *Times Picayune*, July 22, 2010. Others have reported other New Orleans Streets, such as Rampart Street near Canal, as a black people's Canal Street. See Thomas McLoughlin, *Soft Hearts and Hard Times: A Boys Life* (iUniverse, February 27, 2002). The horn player Anthony "Tuba Fats" Lacen reminisces that Dryades Street uptown was once the black people's Canal Street. See Mick Burns, *Keeping the Beat on the Street: The New Orleans Brass Band Renaissance*, (Baton Rouge, La.: LSU Press, February 1, 2008).

37. The data for this paragraph was obtained from http://www.datacenterresearch .org/data-resources/neighborhood-data/district-4/Seventh-Ward/.

38. http://www.datacenterresearch.org/pre-katrina/orleans/7/24/snapshot.html.

39. The data for the following two paragraphs were obtained from http://www.data centerresearch.org/data-resources/neighborhood-data/district-3/Audubon/; http://www.datacenterresearch.org/data-resources/neighborhood-data/district-3/; and http://www.datacenterresearch.org/data-resources/neighborhood-data/district-3/ Freret/.

40. See Richard Campanella, "Gentrification and Its Discontents: Notes from New Orleans," *New Geography*, March 1, 2013, http://www.newgeography.com/content /003526-gentrification-and-its-discontents-notes-new-orleans.

41. I'm not sure if using the term "magic window" for ATM is New Orleans jargon, but my father always referred to it as such. Unfortunately, for most New Orleanians growing up in the city, that magic runs out fast.

42. See Richard Campanella, "What Does It Mean to Be a New Orleanian?" *Times Picayune*, December 8, 2013, http://www.nola.com/opinions/index.ssf/2013/12/ whatdoesitmeantobeanew.html.

43. Kelsey Nowakowski, "Charts Show How Hurricane Katrina Changed New Orleans: Hurricane Katrina Had Lasting Effects on the Physical and Social Makeup of the Big Easy," *National Geographic*, August 29, 2015, http://news.nationalgeographic .com/2015/08/150828-data-points-how-hurricane-katrina-changed-new-orleans/.

3. LIVING DOWN AND OUT IN NEW ORLEANS

1. Charles Bukowski, *Factotum* (Santa Barbara, Calif.: Black Sparrow, 1975).

2. Noam Chomsky and David Barsamian, *Propaganda and the Public Mind: Interviews by David Barsamian* (Chicago: Haymarket, 2015), 181–82.

238 3. Living Down and Out in New Orleans

3. Judith A. Ranta, *Women and Children of the Mills: An Annotated Guide to Nineteenth-Century American Textile Factory Literature*, annotated ed. (Westport, Conn.: Greenwood, 1999).
4. Boris was one of the main characters of Orwell's *Down and Out in Paris and London* who—for better or worse—helped Orwell initially.
5. My childhood home was destroyed during Hurricane Katrina and, at least in physical appearance, is now nothing more than an empty lot. For me, it is still home, a place where I continue to make pilgrimages. I still carry the keys to unlock a front door that is no longer there.
6. These jobs included working in the visitor-services department at the Aquarium of the Americas, delivering food at Vazquez Seafood Restaurant, bartending at Parley's Tavern, delivering food and waiting tables at Café Roma, bouncing at Club 735 on Bourbon Street, and delivering food and waiting tables at Fellini's Restaurant in Mid-City.
7. To her credit, Moira and I discussed this topic of refusing water to customers at length. Although I'm no saint, I argued that all humans have a right to water, free water. I pointed out that water is a universal right. To refuse water to our fellow humans is, in my view, cruel. We argued over this, almost negatively affecting our otherwise good working relationship. Over time, she changed her views. So long as a customer is respectful and appreciative of getting the water, it is no longer seen as getting over. (Moira did always gave water to children, the elderly, and pregnant women.) Moira is a strong New Orleans woman, tough as nails on the outside, and though she may not admit it, she is every bit as tender and caring on the inside.
8. Of course, many people of the rich European, Asia, Latin American, and African cultures tip, and for that, we in the service industry appreciate you.
9. The strategies for surviving down-and-out urban life presented here derive from my personal experiences living in the downtrodden urban underbelly of New Orleans.
10. Data on the cost of goods and services in 1929 Paris was derived from Jeanne Singer-Kerel, *Le cout de la vie à Paris de 1840 à 1954* (Paris: Librairie Armand Colin, 1961).
11. All U.S. dollar amounts are the current price adjusted using the CPI.
12. Katherine Sayre, "New Orleans Area Apartment Rents, Occupancy Rates Continue to Rise," *Times Picayune*, June 14, 2013.

4. BUSKERS, HUSTLERS, AND STREET PERFORMERS

1. Busking is the act of performing services on the street for entertainment, mainly for tips.
2. The New Orleans busker Washboard Brad tells the story of how he once busked in the same spot as Trombone Shorty at the A&P on Royal and Toulouse in the early

1980s. He explains how they had a battle of the bands to gain the interest of the tourists. "We beat them at the time. They were little kids with trombones and horns, but we were adults, so we won. Well, in the end, I guess Trombone Shorty won."

3. Philippe Bourgois found that the informal underground crack economy served as an equal-opportunity employer for poor and working-class Puerto Rican men in Harlem excluded from respectable and living-wage jobs in the formal economy. See Philippe Bourgois, *In Search of Respect: Selling Crack in El Barrio*, 2nd ed. (Cambridge: Cambridge University Press, 2012). Randol Contreras makes a similar argument about Dominican kids from the Bronx robbing drug dealers toward the end of the crack era. See Randol Contreras, *The Stickup Kids: Race, Drugs, Violence, and the American Dream* (Berkeley: University of California Press, 2012).

4. The experienced New Orleans busker Washboard Brad explained in our interview how tourists in the 1980s found New Orleans a strange but liberating place. Once they realized that it was legal and acceptable to drink and dance on the street, they would "relax their sphincter" to his music. He enjoyed looking at "straight white people from, I dunno, somewhere else" get that "magic moment light turned on" when they realized they could sing and dance with the buskers while drinking on the street. "Tourists were uptight but wanted to loosen up." "Now," he says mockingly, "tourists say, 'I'm uptight and I'm proud of it.'" He explains that there is no more of that magic moment; tourists are losing it. He also complains that New Orleans is becoming "white Disneyfied shit" and is now dying. He says, "I'm going to have to go to Havana." I think about how maybe, just maybe, there are no rubrics there.

5. Truman Capote, "Hidden Gardens," in *Music for Chameleons* (1980; New York: Vintage, 1994).

6. Jock Young, *The Exclusive Society: Social Exclusion, Crime, and Difference in Late Modernity* (London: Sage, 1999).

7. Barbara Ehrenreich, *Nickel and Dimed: On (Not) Getting By in America* (New York: Metropolitan, 2001).

8. Jock Young, *The Vertigo of Late Modernity* (London: Sage, 2007).

9. David Gladstone refers to the New Orleans governing tourist regime as the city officials, private-sector investors, and others involved in post-Katrina planning who help anchor the city's growth in tourism and other related industries. See David L. Gladstone, "Event-Based Urbanization and the New Orleans Tourist Regime: A Conceptual Framework for Understanding Structural Change in U.S. Tourist Cities," *Journal of Policy Research in Tourism, Leisure, and Events* 4, no. 3 (2012): 221–48.

10. Geoffrey Edwards and Ryan Edwards, *A.K.A. DOC: The Oral History of a New Orleans Street Musician—James May (a.k.a. "Doc Saxtrum")* (Redwood, N.Y.: Cadence Jazz Books, 2000).

11. The famous sociologist Robert Merton developed the concept of innovator to explain one type of creative response to resolve the contradictions between cultural

goals and institutional means. See Robert Merton, "Social Structure and Anomie," *American Sociological Review* 3, no. 5 (October 1938): 672–82.

12. Jim Flynn, *Sidewalk Saints: Life Portraits of the New Orleans Street Performer Family* (Chicago: Curbside, 2011).

13. Ignatius Jacques Reilly is the protagonist from the great New Orleans novel *Confederacy of Dunces,* by John Kennedy Toole (Baton Rouge: Louisiana State University Press, 1980). Ignatius sold hot dogs near and around the French Quarter, sometimes mocking potential customers for eating unhealthy food while himself scarfing down hot dogs, and, in turn, his profits.

14. Mitch Duneier, *Sidewalk* (New York: Farrar, Straus and Giroux, 2000), 6.

15. Flynn, *Sidewalk Saints.*

16. When students ask what they can do with a degree in sociology, I always tell them that a degree in sociology will make you better at whatever you are going to do in life.

17. Although the local sportswriters in New Orleans seem to have no problem calling the New Orleans Superdome the Mercedes-Benz Superdome, I take objection to private corporations taking over the naming rights of public spaces and buildings. This is especially the case considering that the Mercedes-Benz company used forty thousand Jewish slaves to produce cars and engines to assist the Nazis in the killing of both Jews and Americans, among other groups. According to the historian Alan Clark: "The slaves 'toiled eighteen hours a day; cowering under the lash, sleeping six to a dog kennel eight feet square, starving or freezing to death at the whim of their guards.'" Alan Clark, *Barbarossa: The Russian-German Conflict, 1941–45* (1965; New York: William Morrow, 1985), 321. Daimler, which owned Mercedes-Benz, not only enslaved Jews to build weapons that killed their own people but also made cars Hitler used during the war and developed planes, tanks, and submarine engines to kill Jews, Americans, and other Europeans. Alan Hall, "Revealed: How the Nazis Helped German Companies Bosch, Mercedes, Deutsche Bank, and VW Get VERY Rich Using 300,000 Concentration Camp Slaves," *Daily Mail,* June 20, 2014, http://www.dailymail.co.uk/news/article-2663635/Revealed-How-Nazis-helped-German-companies-Bosch-Mercedes-Deutsche-Bank-VW-VERY-rich-using-slave labor.html. It is possible that Mercedes-Benz participated in the killing of more Jews and Americans than any group that the U.S. government considers terrorists.

18. Gold Man acquired the football he uses in his street performances from fellow busker and New Orleans icon Uncle Louie in 1986. Gold Man knew Uncle Louie, an electrician by trade, from drinking and sometimes working together. Adhering to the informal codes of the busker, Gold Man asked Uncle Louie if he could use the football in his act.

19. "Who Dats" refers to the fans of the New Orleans Saints NFL team. "Who dat" is part of a chant Saints fans say to opposing teams and their fans: "Who dat say dey gonna beat dem Saints. Who dat? Who dat?"

20. See Edward Sutherland, *The Professional Thief* (Chicago: University of Chicago Press, 1937).

5. THE INFORMAL NOCTURNAL ECONOMY OF FRENCHMEN STREET

1. http://www.myneworleans.com/New-Orleans-Magazine/September-2013/Is-Frenchmen-Street-the-next-Bourbon-Street/. "Frenchmen Street is getting more and more popular with tourists, which is leading to questions about its future." In this article, Haley Adams writes: "Often referred to as the 'locals' Bourbon Street,' Frenchmen Street is clearly no longer a locals' secret. The street's famous jazz clubs paired with its wonderfully weird New Orleans character has made Frenchmen Street a favorite suggestion for savvy tourists looking to explore beyond Bourbon Street. Travel magazines and blogs love to include the location as a must-see for visitors."
2. The native of New Orleans provides geographical directions based on proximity to the lake or the river (e.g., "riverbound" and "lakebound").
3. Unless otherwise specified, Shannon Monaghan wrote the poems for this section.
4. "Supplement 72; online content updated on June 29, 2015; CODE OF ORDINANCES— City of NEW ORLEANS, LOUISIANA; Codified through Ordinance No. 26429, adopted May 14, 2015. Note that: This Code of Ordinances and/or any other documents that appear on this site may not reflect the most current legislation adopted by the municipality."
5. I'm loosely borrowing my use of "voodoo" from Jock Young's "Voodoo Criminology and the Numbers Game," in *Cultural Criminology Unleashed*, ed. J. Ferrell et al. (London: GlassHouse, 2004). To an outsider, Voodoo is mysterious and makes little sense. This in no way invalidates the practice of Voodoo.
6. Ben Meyers, "Boil Water Advisory Remains in Effect for New Orleans East Bank," *Times-Picayune*, July 25, 2015.
7. "Hobo gold" refers to the *Crew Change Guide*, which is an underground document that train hoppers and hobos share. It contains detailed information about North American train yards, crew-changing schedules, and maps. Shannon gave me a copy of the most updated *Crew Change Guide*.

6. CITY SQUATTING AND URBAN CAMPING

1. Jonathan Bullington, "French Quarter Transients Face Arrest, and More, on New Orleans Streets," *Times Picayune*, March 30, 2015, http://www.nola.com/crime/index.ssf/2015/03/french_quarter_transients_targ.html.
2. Erving Goffman, "The Moral Career of the Mental Patient," *Psychiatry: Journal for the Study of Interpersonal Processes* 22 (1959): 123–42.
3. David Amsden, "Who Runs the Streets of New Orleans? How a Rich Entrepreneur Persuaded the City to Let Him Create His Own High-Tech Police Force," *New York Times Magazine*, July 30, 2015.

4. Ibid.

5. Nels Anderson explains "main stems" and the concept of Hobohemia: "Every large city has its district into which these homeless types gravitate. In the parlance of the 'road' such a section is known as the 'stem' or the 'main drag.' To the homeless man it is home. . . . Hobohemia is divided into four parts—west, south, north, and east—and no part is more than five minutes from the heart of the Loop. They are all the "stem" as they are also Hobohemia. This four-part concept, Hobohemia, is Chicago to the down-and-out." Nels Anderson, *The Hobo: The Sociology of Homeless Men* (Chicago: University of Chicago Press, 1923).

6. Ibid.

7. Ibid.

8. Laura Maggi, "Toll of Eight Lives in 9th Ward Fire Is Highest in New Orleans in Decades," *Times-Picayune*, December 28, 2010.

9. Gigi refers to Pratchett: "It was octarine, the colour of magic. It was alive and glowing and vibrant and it was the undisputed pigment of the imagination, because wherever it appeared it was a sign that mere matter was a servant of the powers of the magical mind. It was enchantment itself. But Rincewind always thought it looked a sort of greenish-purple." Terry Pratchett, *The Colour of Magic* (New York: St. Martin's Press, 1983).

10. With a "*laissez*" in front of it, "*le bon temps rouler*" becomes the unofficial motto of New Orleans, translating to "let the good times roll."

11. Michelle Krupa, "Census Data Show Large Number of Empty Homes in New Orleans Area," *NOLA.com*, May 25, 2011, http://www.nola.com/politics/index.ssf/2011/05/census_data_show_large_number.

12. Allison Plyer, "Population Loss and Vacant Housing in New Orleans Neighborhoods," *The Data Center Analysis of U.S. Census Bureau, Decennial Census*, February 5, 2011, http://www.datacenterresearch.org/reports_analysis/population-loss-and-vacant-housing/.

13. United States, U.S. Census Bureau, American Factfinder' Factfinder.census.gov (Washington, D.C.: U.S. Dept. of Commerce, Economics and Statistics Administration, U.S. Census Bureau, 2014), http://factfinder.census.gov/faces/tableservices/jsf/pages/productview.xhtml?src=bkmk.

14. Josh Mack, *The Hobo Handbook: A Field Guide to Living by Your Own Rules* (Avon, Mass.: Adams Media, 2011).

15. The Hobo National Convention meets in the town of Britt, Iowa, and is widely believed to be the largest gathering of hobos in the country.

16. Mack, *The Hobo Handbook*, 8–9.

17. Anderson, *The Hobo* (Chicago: University of Chicago Press, 1923).

18. Aliases are used for names of all of the black squatters of New Orleans.

19. Washitah is sometimes also spelled Washitaw.

20. This was also later reported in the *Times Picayune*: Robert McClendon, "Black Nationalist 'Washitah Nation' Claims Bywater House, Changes Locks While Home Is for Sale," *Times Picayune*, February 15, 2016.

21. Robert McClendon, "Read the Washitah Nation Manifesto That Squatters Used to Flummox the NOPD," *Times Picayune*, March 3, 2016.

22. Ibid.

23. Ibid.

24. Sally Richardson, "Abandonment and Adverse Possession," *Houston Law Review* 52, no. 1385.

25. State of Louisiana Court of Appeal, Third Circuit (06-1087), *Caffery Alexander vs. Michael Rene Maddox, et al.* (appeal from the Sixteenth Judicial District Court parish of St. Martin, no. 66857, Honorable Keith Rayne Jules Comeaux, District Judge).

26. Emily Lane, "Jailed 'Washitah Nation' Squatters Refuse to Sign Documents, Offer Names 'Sub Zero,' 'Batman,'" *Times Picayune*, February 17, 2016.

27. Steven Spitzer, "Toward a Marxian Theory of Deviance," *Social Problems* 22 (1975): 638–651.

28. Ibid.

29. Robert McClendon, "Boarding School Graduate's Search for Answers Leads to Washitaw 'Nation,' Jail," *Times Picayune*, March 11, 2016.

30. Ibid.

31. I received full permission to spend a night at the homeless shelter from one of the head managers in charge of the New Orleans Mission. Rest assured, the bed space used and food consumed did not prevent another homeless person from getting a meal or bed space. The daytime manager made sure that my identity as a sociologist conducting research would not be revealed to the other workers or patrons.

32. Or perhaps it was one of Cage's *National Treasure* movies.

33. Richardson, "Abandonment and Adverse Possession."

34. Charles Mills, *The Sociological Imagination* (Oxford: Oxford University Press, 1959).

35. In an interview in May 1996 on the television show *60 Minutes*, Secretary of State Madeleine Albright said, when asked if the United States' social policy on U.S. sanctions against Iraq—which is believed to have led to the death of half a million children—was worth it, she replied, "I think this is a very hard choice, but the price—we think the price is worth it."

7. OCCULTISTS AND SATANISTS

1. On October 2006, New Orleans transplant resident Zackery Bowen, who arrived in the city from Los Angeles, killed, dismembered, and cooked his girlfriend Addie Hall. The couple fell in love the night of Hurricane Katrina and became known in the aftermath of the storm as being flood diehards who had refused to leave the city. Zackery committed suicide, jumping off the roof of the Omni Royal Orleans hotel, with a note in his pocket describing the murder of Addie and where to find her body.

2. Alex Mar, *Witches of America* (New York: Sarah Crichton Books/Farrar, Straus and Giroux, 2015).

3. Many of the people and places in this section and throughout this chapter are given aliases to protect the identity of the research participants. In many cases, the researcher requested that participants in this part of the research project purposely provide fictitious names to protect their identities in case of judicial hostility.

4. As tempting as it is, the name of this local celebrity must be omitted here to protect my sources.

5. The Pentecostal experience of getting the Holy Ghost often happens to born-again believers who experience divine supernatural gifts like, for example, speaking in tongues. See Peter Marina, *Getting the Holy Ghost: Urban Ethnography in a Brooklyn Pentecostal Tongue-Speaking Church* (Lanham, Md.: Lexington, 2014).

6. Ben Estes and Ben Myers, "Police Make Gruesome Discovery in Slidell Home: Human and Animal Remains," *Times Picayune*, March 31, 2016, http://www.nola.com/crime/index.ssf/2016/03/police_make_gruesome_discovery.html.

7. Jim Mustian, "Bizarre Facebook Post on Collecting Human Remains Leads Police to Raid Witch's Mid-City Home, Find Bones, Teeth," *New Orleans Advocate*, March 29, 2016, http://www.theneworleansadvocate.com/news/15297574-65/in-bone-trafficking-probe-state-authorities-find-human-remains-after-raiding-mid-city-home.

8. 2011 Louisiana Laws. Revised Statutes TITLE 8—Cemeteries RS 8:653—Opening graves; stealing body; receiving same.

9. Elizabeth Crisp, "After Raid of Witch's New Orleans Home Uncovers Bones, Teeth, State Lawmakers Eye Stiffer Penalties for Trafficking Human Remains," *New Orleans Advocate*, April 5, 2016, http://theadvocate.com/news/15397249-123/after-case-of-self-styled-witch-draws-attention-in-new-orleans-state-lawmakers-eye-bill-to-create-st.

10. David Grunfeld, "Little Freddie King Celebrates Seventy-Fifth Birthday at BJ's Lounge: Photos and Video," *Times Picayune*, July 19, 2015, http://www.nola.com/music/index.ssf/2015/07/little_freddie_kings_75th_birt.html.

11. The authors of Gumbo Ya-Ya refer to St. Roch Cemetery as one of the most unusual cemeteries in New Orleans. Robert Tallant and Lyle Saxon, *Gumbo Ya-Ya: A Collection of Louisiana Folk Tales*, 1st pbk. ed. (Gretna, La.: Pelican, 1987).

12. Ibid.

13. Although Damian argues that these Afro-Brazilian religious practices emerged from favelas, it is most probable that these religious practices emerged out of slave communities, since these religious practices existed before the development of favelas.

8. GENTRIFICATION AND VIOLENT CULTURAL RESISTANCE

1. "Greater New Orleans Community Data Center," http://www.datacenterresearch.org/pre-katrina/orleans/7/24/snapshot.html.

2. Todd Price, "Long-Empty St. Roch Market Selects Vendors and Readies to Return," *Times Picayune*, December 10, 2014, http://www.nola.com/dining/index.ssf/2014/12/long-empty_st_roch_market_sele.html.

3. Ibid.

4. Jaquetta White, "Vandalization of New St. Roch Market Reflects Community's Dissatisfaction, Disappointment in Finished Product, Residents Say," *New Orleans Advocate*, May 2, 2015.

5. Ibid.

6. Richard Webster, "St. Roch: Gentrification Ground Zero in New Orleans," *Times Picayune*, June 9, 2015.

7. Ibid.

8. This was also the racist basis for *Brown v. Board of Education*, which symbolically desegregated public schools. Supreme Court's idea was that white people made black schools better, not the other way around. See Mamadou Chinyelu, *Harlem Ain't Nothin' but a Third World Country: The Global Economy, Empowerment Zones, and the Colonial Status of Africans in America* (New York: Mustard Seed, 1999).

9. Data from this section was obtained from multiple sources, including the U.S. Census Data Research Center (http://www.datacenterresearch.org/data-resources/katrina/facts-for-impact/) and the *Times Picayune*.

10. David Gladstone, "Hard Times in the Big Easy: Disaster Recovery in a Tourism-Dependent City," unpublished paper presented at the fifty-fifth annual Association of Collegiate Schools of Planning (ACSP) in Houston (2015).

11. David L. Gladstone, "Event-Based Urbanization and the New Orleans Tourist Regime: A Conceptual Framework for Understanding Structural Change in U.S. Tourist Cities," *Journal of Policy Research in Tourism, Leisure and Events* 4 (2012): 221–248.

12. Gladstone, "Hard Times in the Big Easy." He received this data from "World Bank, 2014. Literacy rate, adult total (percent of people ages 15 and above)," http://data.worldbank.org/indicator/SE.ADT.LITR.ZS.

13. Peter Marina (with Vern Baxter), "Cultural Meaning and Hip-Hop Fashion in the African-American Male Youth Subculture of New Orleans," *Journal of Youth Studies* 11, no. 2 (2008): 93–113.

14. David L. Gladstone, "Event-Based Urbanization"; James Dao and N. R. Kleinfeld, "Conditions in New Orleans Still Dire—Pumping May Take Months," *New York Times*, September 3, 2005.

15. U.S. Department of Justice, "U.S. Department of Justice, Prison and Jail Inmates at Mid-Year, 2005," Washington, D.C. (2006).

16. Ibid.

17. E. Ann Carson, "Prisoners in 2013 and States of Incarceration: The Global Context," U.S. Department of Justice, http://www.prisonpolicy.org/global/.

18. Nearly 3 percent of all black men in the United States are behind bars, compared to 0.5 percent of all white men. Black men who were eighteen or nineteen

years old were nine times more likely to be imprisoned than their white peers in 2013.

19. Sharon Zukin, "Gentrification: Culture and Capital in the Urban Core," *Annual Review of Sociology* 13 (1987): 129–147.

20. Richard Campanella, "Gentrification and Its Discontents: Notes from New Orleans," *New Geography*, March 1, 2013, http://www.newgeography.com/content/003526-gentrification-and-its-discontents-notes-new-orleans.

21. Gladstone argues that in the 1970s "neighborhoods in or adjacent to the city's tourism zones were initially gentrified by gay waiters and other tourism workers in French Quarter restaurants, in much the same way that aspiring artists were some of the first gentrifiers in Soho, Chelsea, and other Manhattan neighborhoods." Gladstone, "Event-Based Urbanization and the New Orleans Tourist Regime"; Gladstone, "Hard Times in the Big Easy."

22. Campanella, "Gentrification and Its Discontents."

23. Gladstone, "Event-Based Urbanization and the New Orleans Tourist Regime"; Gladstone, "Hard Times in the Big Easy."

24. Lovell Beaulieu, "Gentrification: The New Segregation?" *New Orleans Tribune*, July 2, 2012.

25. John Arena, *Driven from New Orleans: How Nonprofits Betray Public Housing and Promote Privatization* (Minneapolis: University of Minnesota Press, 2012).

26. Ibid.

27. City of New Orleans, Uniform Crime Reports, "New Orleans U.C.R. Reportable Crimes 2015 Compared to 2014," http://www.nola.gov/getattachment/86272c26-cb13-4573-8daa-8f19c4cfcf47/2015-Year-End-UCR-Citywide/.

28. "New Orleans Ends 2015 with 164 Murders, Up After Two-Year Drop," *Times Picayune,* January 1, 2016.

29. Gladstone, "Event-Based Urbanization and the New Orleans Tourist Regime"; Gladstone, "Hard Times in the Big Easy."

30. Robert McClendon, "Where Will Working Poor Live in Future New Orleans, If Gentrification Continues?" *Times Picayune*, July 30, 2015.

31. Peter Tatian, a senior research associate in the Urban Institute's Metropolitan Housing and Communities Policy Center explains: "To be sure, gentrification brings with it many benefits—new investment in the neighborhood, increased homeownership, appreciation of property values, and new residents who can add vitality to the community. But as housing costs rise from increasing demand, existing residents can find it more difficult to afford to remain in the neighborhood. There are steps that can be taken to deal with these concerns, but it is vital that the city government and neighborhood residents work together toward balancing those issues." See Allison Plyer and Denice Warren Ross, "Change in the Irish Channel: Thirty Years of Data About a Historic New Orleans Neighborhood," *Greater New Orleans Community Data Center*, http://www.datacenterresearch.org/prekatrina/articles/IrishChannel.html.

32. I personally remember as a teacher in New Orleans, at the now demolished John F. Kennedy High School, students dividing the school into informal Seventh Ward "Hard Head" and Ninth Ward "Skull Crusher" sections.

33. The longtime New Orleans resident and business owner Armand Charbonnet remembers the "big oak trees and azalea gardens" on Claiborne Avenue that he refers to as the "black people's Canal Street." See Lolis Elie, "Planners Push to Tear out Elevated I-10 Over Claiborne," *Times Picayune*, July 11, 2009.

34. The result of this project led to the isolation of Red Hook from the rest of Brooklyn and many of its now economically vibrant neighborhoods. Just as wealthier residents of the French Quarter were able to move the Riverfront Expressway project to the lower-income and economically more powerless neighborhoods along Claiborne Avenue, affluent Brooklyn Heights residents pushed for Gowanus Expressway project to impact most negatively the working-class Red Hook neighborhood. On a side note, Red Hook today is a wealthy and nearly fully gentrified community of mostly wealthy middle- and upper-middle-class mono-European Americans. Richard Nalley, "Unsinkable Red Hook: Brooklyn's Forgotten Waterfront May Finally Be Getting Some Love," *New York Observer*, May 13, 2015, http://observer.com/2015/05/unsinkable-red-hook/.

35. Elie, "Planners Push to Tear out Elevated I-10 Over Claiborne."

36. From a video discussing the importance of the New Orleans Claiborne Corridor, https://www.youtube.com/watch?v=f29KrbqcCxU.

37. Gladstone, "Hard Times in the Big Easy."

38. Lyndon Jones, "Many Residents Leery of 'Claiborne Corridor' Study," *New Orleans Tribune*, May 2016, http://www.theneworleanstribune.com/main/many-residents-leery-of-claiborne-corridor-study/.

39. From a video discussing the importance of the New Orleans Claiborne Corridor, https://www.youtube.com/watch?v=f29KrbqcCxU.

40. Peter Moskowitz, "Destroy and Rebuild: A Q&A with One of New Orleans' Biggest Developers," *True Stories*, February 2015, http://truestories.gawker.com/destroy-and-rebuild-a-q-a-with-one-of-new-orleans-bigg-1684973590.

41. Lance Freeman, *There Goes the 'Hood: Views of Gentrification from the Ground Up*, (Philadelphia: Temple University Press, 2006).

42. Arena, *Driven from New Orleans*.

9. HIPSTER WONDERLAND

1. The Supplemental Nutrition Assistance Program (SNAP) supplies eligible low-income families and individuals with nutrition assistance.

2. Hipsters often hold "song shares" in public and private venues where they take turns singing songs and reciting poetry about things they find interesting, usually their own lives. A pop-up café is a restaurant backyard speakeasy.

3. See J. Lea and Jock Young, "Relative Deprivation," in *Criminological Perspectives: A Reader*, ed. J. Muncie, E. MacLaughlin and M. Langan (London: Sage, 1996), 136.
4. Jeannie Haubert, *Rethinking Disaster Recovery: A Hurricane Katrina Retrospective*, (Lanham, Md.: Lexington, 2015).
5. See Jock Young, *The Criminological Imagination* (Cambridge: Polity, 2011).

10. BRASS BANDS AND SECOND LINES

1. Louis Armstrong's "When the Saints Go Marching in" inspired this sentence. The exact origins of this Christian gospel hymn are unknown.
2. "Social Aid and Pleasure Club" is a common part of parading clubs' names in New Orleans. Social aid and pleasure clubs—often abbreviated "S&P" or "SA&PC"—were originally postemancipation service organizations that provided assistance to black people on a variety of matters, including health and unemployment and funeral insurance. Today, the social aid and pleasure clubs of New Orleans organize the city's second lines, like the one discussed in this chapter, providing members of the black community, as well as all residents of New Orleans, a chance to celebrate the culture of the city on its neighborhood streets. See David Kunian, "'A Positive, Cultural Thing': Social Aid and Pleasure Clubs Continue a Centuries-Old Tradition of 'Second-Line' Parades," *Gambit*, February 6, 2007.
3. The legendary New Orleans musician Dr. John describes the meaning of the Mardi Gras Indian phrase "Tu way pocky way" in the following passage: "To keep an eye out for other tribes roaming around the city on Mardi Gras day, a tribe sent out the spy boy as a runner. If one spy boy bumped into another spy boy from some other tribe, they'd have an exchange like so: 'Om bah way,' the first spy boy might say. And the other spy boy, if everything was cool and there wasn't going to be no confrontation, answered: 'Tu way pocky way.' This talk was the Indian's own Creole language, part French, part Spanish, part Choctaw, part Yoruba, and part mystery to an outsider like me. What the first one said basically was, 'Where yatt, bro?' or the like. And the second one said, 'Everything's oaks and herbs'—which means everything's cool because they had smoked lots of herbs. If the second one responded 'No om bah way,' then y'all had problems, and a challenge by way of each and other had been issued." See Mac Rebennack (a.k.a. Dr. John) and Jack Rummel, *Under a Hoodoo Moon: The Life of the Night Tripper*, repr. ed. (New York: St. Martin's Griffin, 1995).
4. I'm reminded of what Tamara Jackson, president of the Social Aid and Pleasure Club Task Force and the VIP Ladies and Kids of New Orleans, stated at a community meeting in New Orleans on gentrification. Ms. Jackson responded to an outside transplant who asked if she (indicating white people not from New Orleans) was welcome to attend the second lines of the city. Ms. Jackson responded as follows: "I want to ask you a question: At your attendance at these events, do you feel safe? Do

you feel the rhythm and the love and the spirit? 'Cause everything from the club's attire has meaning, some stops are traditional and historically prevalent for that organization. Everything has meaning. And what hurts us is when people come and they miss the whole point. When it become pointless to you, then I'm going to ask you not to come. But as long as you value the experience, and you can appreciate what we do. Take a moment when we stop at those watering holes, ask if you have some questions. They have members that'll be willing to share with you if you have some questions about why the club change, why we stopped here, why was the slow procession doing the uh—'cause y'all like to jump and boogie. When the music gets slow, honey y'all look like y'all be sad standing to the side. But it's okay, because everything has meaning and if you want to know, ask. I enjoy it, We want y'all to come all the time."

5. Travis "Trumpet Black" Hill died on May 4, 2015, about a week after a minor dental procedure led to an infection that quickly spread quickly to his heart. He was a much-loved local musician who played with the New Birth Brass Band and his own band Heart Attacks, among others. Prior to picking up a trombone and becoming a New Orleans musician, Hill served a nine-year prison sentence for armed robbery. Once released, he did as his song says, used trumpets and not guns while playing music in the city and helping at-risk kids. See Allison Fensterstock, "Travis 'Trumpet Black' Hill Laid to Rest in New Orleans Saturday (May 23) with Music, Tributes," *Times Picayune*, May 24, 2015.